Underwater Acoustic Sensor Networks

OTHER TELECOMMUNICATIONS BOOKS FROM AUERBACH

AUERBACH PUBLICATIONS
www.auerbach-publications.com
To Order Call: 1-800-272-7737 • Fax: 1-800-374-3401
E-mail: orders@crcpress.com

Underwater Acoustic Sensor Networks

Edited by YANG XIAO

CRC Press
Taylor & Francis Group
Boca Raton   London   New York

CRC Press is an imprint of the
Taylor & Francis Group, an **informa** business

AN AUERBACH BOOK

Auerbach Publications
Taylor & Francis Group
6000 Broken Sound Parkway NW, Suite 300
Boca Raton, FL 33487-2742

Library of Congress Cataloging-in-Publication Data

Underwater acoustic sensor networks / editor, Yang Xiao.
    p. cm.
Includes bibliographical references and index.
ISBN 978-1-4200-6711-8 (hardcover : alk. paper)
1. Underwater acoustic telemetry. 2. Wireless sensor networks. I. Xiao, Yang, 1966-

TK5103.52.U54 2010
681'.2--dc22   2009053599

Visit the Taylor & Francis Web site at
http://www.taylorandfrancis.com

and the Auerbach Web site at
http://www.auerbach-publications.com

Contents

SECTION III   FAULT TOLERANCE AND TIME SYNCHRONIZATION

SECTION IV   MEDIUM ACCESS CONTROL

SECTION V   SOFTWARE, HARDWARE, AND CHANNEL MODELING

Preface

Underwater acoustic sensor networks are different from terrestrial radio-based sensor networks in many aspects such as propagation delay, bandwidth, transmit energy, multi-path effect, etc. For example, acoustic signals have a propagation speed that is around five orders of magnitude slower than radio signals. Therefore, many algorithms and protocols from application layer to physical layer need to be re-designed.

This book investigates fundamental aspects of underwater acoustic communications. It presents a collection of recent advances in underwater sensor network areas. It is a contribution of many prominent researchers working in the sensor networks field around the world. The book contains 12 chapters that are divided into five sections, and it covers a wide spectrum of underwater sensor networks including, but not limited to, research challenges, clustering, topology control, routing, fault tolerance, time synchronization, medium access control, software, hardware, channel modeling, etc. Nevertheless, I believe this book will be a good reference for researchers, practitioners, and students who are interested in the research, development, design, and implementation of underwater sensor networks.

This book was made possible by the great efforts of the contributors and publisher. I am indebted to the contributors, who have sacrificed days and nights to put together these chapters for our readers. We would like to thank our publisher. Without their encouragement and quality work, we could not have this book.

Yang Xiao
Department of Computer Science
University of Alabama
Tuscaloosa, Alabama, USA

Acknowledgment

This work is supported in part by the U.S. National Science Foundation (NSF) under the grant numbers CCF-0829827, CNS-0716211, and CNS-0737325.

About the Editor

Yang Xiao worked in industry as a MAC (Medium Access Control) architect involving the IEEE 802.11 standard enhancement work before he joined the Department of Computer Science at the University of Memphis in 2002. He is currently with the Department of Computer Science at the University of Alabama. He was a voting member of the IEEE 802.11 Working Group from 2001 to 2004. He is an IEEE Senior Member and a member of the American Telemedicine Association. Dr. Xiao currently serves as editor-in-chief for the *International Journal of Security and Networks* (IJSN), *International Journal of Sensor Networks* (IJSNet), and *International Journal of Telemedicine and Applications* (IJTA). He serves as a referee/reviewer for many funding agencies, as well as a panelist for NSF, and he is a member of the Canada Foundation for Innovation's (CFI's) Telecommunications expert committee. He has served on TPC for more than 100 conferences, including INFOCOM, ICDCS, MOBIHOC, ICC, GLOBECOM, WCNC, etc. He serves as an associate editor for several journals, e.g., *IEEE Transactions on Vehicular Technology*. His research areas are security, telemedicine, sensor networks, and wireless networks. He has published more than 300 papers in major journals, refereed conference proceedings, and written book chapters related to these research areas. Dr. Xiao's research has been supported by the U.S. National Science Foundation (NSF), U.S. Army Research, Fleet and Industrial Supply Center San Diego (FISCSD), and the University of Alabama's Research Grants Committee.

Contributors

S. M. Nazrul Alam
Department of Computer Science
Cornell University
Ithaca, New York

Praveer Bahri
Department of Computer Science
University of Alabama
Tuscaloosa, Alabama

James Byron
Department of Computer Engineering
Rochester Institute of Technology
Rochester, New York

Jiming Chen
State Key Laboratory of Industrial
    Control Technology
Institute of Industrial Process Control
Zhejiang University
Hangzhou, China

Min Chen
Department of Electrical and
    Computer Engineering
University of British Columbia
Vancouver, British Columbia, Canada

Xiuzhen Cheng
Department of Computer Science
George Washington University
Washington, DC

Mandar Chitre
National University of Singapore
Singapore

Michael Galloway
Department of Computer Science
University of Alabama
Tuscaloosa, Alabama

Amir Aminzadeh Gohari
Department of Electrical and
    Computer Engineering
University of California, Santa
    Barbara
Santa Barbara, California

Zygmunt J. Haas
Wireless Networks Laboratory
School of Electrical and Computer
    Engineering
Cornell University
Ithaca, New York

Fei Hu
Department of ECE
University of Alabama
Tuscaloosa, Alabama

Scott C.-H. Huang
Computer Science Department
City University of Hong Kong
Kowloon, Hong Kong

Raja Jurdak
Computer Science and Informatics
   Centre
University College Dublin
Belfield, Ireland

Madhulika Kamboj
Department of Computer Science
University of Alabama
Tuscaloosa, Alabama

Peter King
Engineering and Applied Science
Memorial University of
   Newfoundland
St. John's, Newfoundland, Canada

Pius W. Q. Lee
Networking Protocols Department
Institute for Infocomm Research (I²R)
Agency for Science, Technology and
   Research (A*STAR)
Singapore

Cheng Li
Engineering and Applied Science
Memorial University of
   Newfoundland
St. John's, Newfoundland, Canada

Qilian Liang
Department of Electrical Engineering
University of Texas at Arlington
Arlington, Texas

Tommaso Melodia
Department of Electrical Engineering
State University of New York at
   Buffalo
Buffalo, New York

Steven Mokey
Department of Computer Engineering
Rochester Institute of Technology
Rochester, New York

Mehul Motani
National University of Singapore
Singapore

Dario Pompili
Department of Electrical and
   Computer Engineering
Rutgers, State University of New Jersey
Piscataway, New Jersey

Volkan Rodoplu
Department of Electrical and
   Computer Engineering
University of California, Santa
   Barbara
Santa Barbara, California

Andrew Sackett
Department of ECE
University of Alabama
Tuscaloosa, Alabama

Winston K. G. Seah
Networking Protocols Department
Institute for Infocomm Research
   (I²R)
Agency for Science, Technology and
   Research (A*STAR)
Singapore

Shiraz Shahabudeen
National University of Singapore
Singapore

Peng Shao
Department of Computer Science
University of Alabama
Tuscaloosa, Alabama

Youxian Sun
State Key Laboratory of Industrial
   Control Technology
Institute of Industrial Process Control
Zhejiang University
Hangzhou, China

H. P. Tan
Networking Protocols Department
Institute for Infocomm Research (I²R)
Agency for Science, Technology and
    Research (A*STAR)
Singapore

Paul Tilghman
Department of Computer
    Engineering
Rochester Institute of Technology
Rochester, New York

Ramachandran Vekatesan
Engineering and Applied Science
Memorial University of
    Newfoundland
St. John's, Newfoundland, Canada

Yang Xiao
Department of Computer Science
University of Alabama
Tuscaloosa, Alabama

Songqing Yue
Department of Computer Science
University of Alabama
Tuscaloosa, Alabama

Jianhui Zhang
College of Computer
Hangzhou Dianzi University
Hangzhou, P.R. China

Xihui Zhang
Department of Computer Information
    Systems
University of North Alabama
Florence, Alabama

Yanping Zhang
Department of Computer Science
University of Alabama
Tuscaloosa, Alabama

Liang Zhao
Department of Electrical Engineering
University of Texas at Arlington
Arlington, Texas

RESEARCH CHALLENGES AND CLUSTERING

I

Chapter 1

Research Challenges in Communication Protocol Design for Underwater Sensor Networks

Dario Pompili and Tommaso Melodia

Contents

Underwater networks of sensors have the potential to enhance our ability to observe and predict the ocean by enabling many applications such as oceanographic data collection, pollution monitoring, offshore exploration, disaster prevention, assisted navigation, and tactical surveillance. Underwater acoustic networking is the enabling technology for these applications. In this chapter, fundamental key aspects of underwater acoustic communications are investigated, and architectures for two-dimensional (2D) and three-dimensional (3D) underwater sensor networks are proposed. A detailed overview of the current acoustic communication solutions for medium access control, network, and transport layer protocols is given and open research issues for protocol design are discussed.

1.1 Introduction

Underwater networks of sensors have the potential to enable unexplored applications and to enhance our ability to observe and predict the ocean. Unmanned or Autonomous Underwater Vehicles (UUVs, AUVs), equipped with underwater sensors, are also envisioned to find application in exploration of natural undersea resources and gathering of scientific data in collaborative monitoring missions. These potential applications will be made viable by enabling communications among underwater devices. Underwater Acoustic Sensor Networks (UW-ASNs) will consist of sensors and vehicles deployed underwater and networked via acoustic links to perform collaborative monitoring tasks.

Underwater acoustic sensor networks enable a broad range of applications, including

- *Ocean Sampling Networks.* Networks of sensors and AUVs can perform synoptic, cooperative, adaptive sampling of the 3D coastal ocean environment.
- *Environmental Monitoring.* UW-ASNs can perform pollution monitoring (chemical, biological, and nuclear), ocean current and wind monitoring, and biological monitoring such as tracking of fish or micro-organisms. Also, UW-ASNs can improve weather forecasting, detect climate change, and understand and predict the effect of human activities on marine ecosystems.
- *Undersea Explorations.* Underwater sensor networks can help detect underwater oilfields or reservoirs, determine routes for laying undersea cables, and assist in exploration for valuable minerals.
- *Disaster Prevention.* Sensor networks that measure seismic activity from remote locations can provide *tsunami* warnings to coastal areas, or study the effects of submarine earthquakes (*seaquakes*).
- *Seismic Monitoring.* Frequent seismic monitoring is of great importance in oil extraction from underwater fields to assess field performance. Underwater sensor networks would allow reservoir management approaches.
- *Equipment Monitoring.* Sensor networks would enable remote control and temporary monitoring of expensive equipment immediately after the deployment, to assess deployment failures in the initial operation or to detect problems.

■ *Assisted Navigation.* Sensors can be used to identify hazards on the seabed, locate dangerous rocks or shoals in shallow waters, mooring positions, submerged wrecks, and to perform bathymetry profiling.

■ *Distributed Tactical Surveillance.* AUVs and fixed underwater sensors can collaboratively monitor areas for *surveillance, reconnaissance, targeting*, and *intrusion detection.*

■ *Mine Reconnaissance.* The simultaneous operation of multiple AUVs with acoustic and optical sensors can be used to perform rapid environmental assessment and detect mine-like objects.

Acoustic communications are the typical physical layer technology in underwater networks. In fact, radio waves propagate at long distances through conductive salty water only at extra-low frequencies (30–300 Hz), which require large antennae and high transmission power. Optical waves do not suffer from such high attenuation but are affected by scattering. Furthermore, transmitting optical signals requires high precision in pointing the narrow laser beams. Thus, links in underwater networks are typically based on *acoustic wireless communications.*[1]

The traditional approach for *ocean-bottom* or *ocean-column* monitoring is to deploy underwater sensors that record data during the monitoring mission, and then recover the instruments.[2] This approach has the following disadvantages:

■ *No real-time monitoring.* The recorded data cannot be accessed until the instruments are recovered, which may happen several months after the start of the monitoring mission. This is critical, especially in surveillance or in environmental monitoring applications such as seismic monitoring.

■ *No on-line system reconfiguration.* Interaction between onshore control systems and the monitoring instruments is not possible. This impedes any adaptive tuning of the instruments, nor is it possible to reconfigure the system after particular events occur.

■ *No failure detection.* If *failures* or *misconfigurations* occur, it may not be possible to detect them before the instruments are recovered. This can easily lead to the complete failure of a monitoring mission.

■ *Limited storage capacity.* The amount of data that can be recorded during the monitoring mission by every sensor is limited by the capacity of the onboard storage devices (memories, hard disks).

These disadvantages can be overcome by connecting untethered underwater instruments by means of wireless links that rely on acoustic communications. Although there exist many recently developed network protocols for wireless sensor networks, the unique characteristics of the underwater acoustic communication channel, such as limited bandwidth capacity and high and variable propagation delays,[2] require very efficient and reliable new data communication protocols.

Major challenges in the design of underwater acoustic networks are

- The available bandwidth is severely limited.
- The underwater channel is impaired because of multipath and fading.
- Propagation delay is five orders of magnitude higher than in Radio Frequency (RF) terrestrial channels, and is variable.
- High bit error rates and temporary losses of connectivity (shadow zones) can be experienced.
- Underwater sensors are characterized by high cost because of a small relative number of suppliers (i.e., not much economy of scale).
- Battery power is limited and usually batteries cannot be recharged.
- Underwater sensors are prone to failures because of fouling and corrosion.

In this survey, we discuss the factors that influence protocol design for underwater sensor networks. The remainder of this chapter is organized as follows. In Sections 1.2 and 1.3 we introduce the main design challenges and the reference communication architectures, respectively, of underwater acoustic networks. In Sections 1.4, 1.5, and 1.6 we discuss medium access control (MAC), network, and transport layer issues in underwater sensor networks, respectively. Finally, in Section 1.7 we draw the main conclusions.

1.2 Design Challenges

In this section, we itemize the main differences between terrestrial and underwater sensor networks, detail the key challenges in underwater communications that influence protocol development, and give motivations for a cross-layer design approach to improve the efficiency of the communication process in the challenging underwater environment.

1.2.1 Differences with Terrestrial Sensor Networks

The main differences between terrestrial and underwater sensor networks can be outlined as follows:

- *Cost.* While terrestrial sensor nodes are expected to become increasingly inexpensive, underwater sensors are expensive devices. This is especially due to the more complex underwater transceivers and to the hardware protection needed in the extreme underwater environment. Also, because of the low economy of scale caused by a small relative number of suppliers, underwater sensors are characterized by high cost.
- *Deployment.* While terrestrial sensor networks are densely deployed, in underwater the deployment is generally more sparse.

■ *Power.* The power needed for acoustic underwater communications is higher than in terrestrial radio communications because of the different physical layer technology (acoustic vs. RF waves), the higher distances, and the more complex signal processing techniques implemented at the receivers to compensate for the impairments of the channel.

■ *Memory.* While terrestrial sensor nodes have very limited storage capacity, underwater sensors may need to be able to do some data caching as the underwater channel may be intermittent.

■ *Spatial Correlation.* While the readings from terrestrial sensors are often correlated, this is more unlikely to happen in underwater networks due to the greater distance among sensors.

1.2.2 Underwater Sensors

The typical internal architecture of an underwater sensor is shown in Figure 1.1. It consists of a main controller/CPU, which is interfaced with an oceanographic instrument or sensor through a sensor interface circuitry. The controller receives data from the sensor and it can store it in the onboard memory, process it, and send it to other network devices by controlling the acoustic modem. The electronics are usually mounted on a frame that is protected by a PVC housing. Sometimes all sensor components are protected by bottom-mounted instrument frames that are designed to permit azimuthally omnidirectional acoustic communications, and protect sensors and modems from the potential impact of trawling gear, especially in areas subjected to fishing activities. The protecting frame should be designed so as to deflect trawling gear on impact, by housing all components beneath a low-profile pyramidal frame.[3]

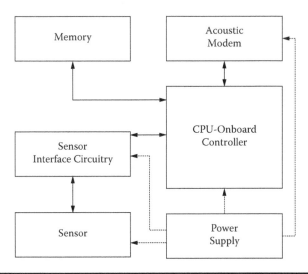

Figure 1.1   Internal organization of an underwater sensor node.

Underwater sensors include sensors to measure the quality of water and to study its characteristics such as temperature, density, salinity (interferometric and refractometric sensors), acidity, chemicals, conductivity, pH (magnetoelastic sensors), oxygen (Clark-type electrode), hydrogen, dissolved methane gas (METS), and turbidity. Disposable sensors exist that detect ricin, the highly poisonous protein found in castor beans and thought to be a potential terrorism agent. DNA microarrays can be used to monitor both abundance and activity level variations among natural microbial populations. Other existing underwater sensors include hydrothermal sulfide silicate, voltammetric sensors for spectrophotometry, gold-amalgam electrode sensors for sediment measurements of metal ions (*ion-selective analysis*), and amperometric microsensors for H_2S measurements for studies of anoxygenic photosynthesis, sulfide oxidation, and sulfate reduction of sediments. In addition, force/torque sensors for underwater applications requiring simultaneous measurements of several forces and moments have also been developed, as well as quantum sensors to measure light radiation and sensors for measurements of harmful algal blooms.

1.2.3 Factors Influencing Underwater Protocol Design

In this section we analyze the main factors in Underwater Acoustic (UW-A) communications that affect the design of protocols at different communication layers. Acoustic communications in the underwater environment are mainly influenced by *transmission loss, noise, multipath, Doppler spread,* and *high and variable propagation delay.* All these factors determine the *temporal and spatial variability* of the acoustic channel, and make the available bandwidth of the underwater acoustic channel limited and dramatically dependent on both range and frequency. Long-range systems that operate over several tens of kilometers may have a bandwidth of only a few kHz, while a short-range system operating over several tens of meters may have more than a hundred kHz of bandwidth. In both cases, these factors lead to low bit rate[4] in the order of tens of kbps for existing devices.

Underwater acoustic communication links can be classified according to their range as *very long, long, medium, short,* and *very short* links.[1] Table 1.1 shows typical

Table 1.1 Typical Bandwidths of Underwater Channel

	Range (km)	Bandwidth (kHz)
Very long	1000	<1
Long	10–100	2–5
Medium	1–10	≈10
Short	0.1–1	20–50
Very short	<0.1	>100

bandwidths of the underwater acoustic channel for different ranges. Acoustic links are also roughly classified as *vertical* and *horizontal*, according to the direction of the sound ray with respect to the ocean bottom. As will be discussed later, their propagation characteristics differ considerably, especially with respect to time dispersion, multipath spreads, and delay variance. In the following, as usually done in oceanic literature, *shallow water* refers to water with depth lower than 100 m, while *deep water* is used for deeper oceans. Hereafter we briefly analyze the factors that influence acoustic communications in order to state the challenges posed by the underwater channels for sensor networking. These include

> *Transmission loss.* The underwater transmission loss describes how the acoustic intensity decreases as an acoustic pressure wave propagates outwards from a sound source. The transmission loss $TL(d, f)$ [dB] that a narrow-band acoustic signal centered at frequency f [kHz] experiences along a distance d [m] can be described by the Urick propagation model,[5] $TL(d, f) = \chi \cdot Log(d) + \alpha(f) \cdot d + A$. The first term accounts for *geometric spreading*, which refers to the spreading of sound energy as a result of the expansion of the wavefronts. It increases with the propagation distance and is independent of frequency. There are two common kinds of geometric spreading: *spherical* (omnidirectional point source, spreading coefficient $\chi = 20$), which characterizes deep water communications, and *cylindrical* (horizontal radiation only, spreading coefficient $\chi = 10$), which characterizes shallow water communications. Inbetween cases show a spreading coefficient χ in the interval $(10, 20)$, depending on water depth and link length. The second term accounts for *medium absorption*, where $\alpha(f)$ [dB/m] represents an absorption coefficient that describes the dependency of the transmission loss on the frequency band, as shown in Figure 1.2. Finally, the last term, expressed by the quantity A [dB], is the so-called *transmission anomaly*, and accounts for the degradation of the acoustic intensity caused by multiple path propagation, refraction, diffraction, and scattering of sound caused by particulates, bubbles, and plankton within the water column. Its value is higher for shallow-water horizontal links (up to 10 dB), which are more affected by multipath.[5] More details can be found in the literature.[6,7]

> *Noise.* It can be classified as *man-made noise* and *ambient noise*. The former is mainly caused by machinery noise (pumps, reduction gears, power plants), and shipping activity (hull fouling, animal life on hull, cavitation), while the latter is related to hydrodynamics (movement of water including tides, current, storms, wind, and rain), and to seismic and biological phenomena. The unique *"V" structure* of the underwater acoustic noise p.s.d. (which has a minimum of 20 $dB_{re\,\mu Pa}$/Hz at about 40 kHz), depicted in Figure 1.3, makes *non-trivial* the choice of the bandwidth. Interestingly, in acoustic communication transmissions, when the central frequency is low, e.g., $f_0 = 10$ kHz, a higher relative signal-to-noise ratio *SNR* is achieved with a narrow bandwidth (e.g., $B = 3$ as opposed to 9 kHz); conversely, when the central frequency

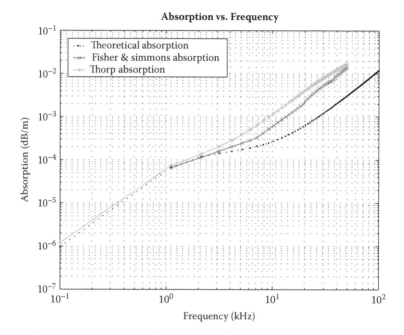

Figure 1.2 Theoretical, Fisher and Simon's, and Thorp's medium absorption coefficient $\alpha(f)$ **versus frequency** $f \in [10^{-1}, 10^2]$ **kHz.**

is high, e.g., $f_0 = 100$ kHz, a higher relative *SNR* is achieved with a wide bandwidth (e.g., $B = 90$ as opposed to 30 kHz). This implies that if a high central frequency is used, a large bandwidth can be exploited for communication, although a high transmit power would be needed to compensate for the higher transmission loss. Acoustic communication solutions tailored for the underwater environment should take into account this unique effect, which is caused by the peculiar "V" structure of the noise p.s.d., and by the fact that the difference between the slopes of the noise and transmission loss decreases with increasing central frequency (e.g., positive for low frequencies and negative for high ones).

Multipath. Multipath propagation may be responsible for severe degradation of the acoustic communication signal, since it generates intersymbol interference (ISI). The multipath geometry depends on the link configuration. Vertical channels are characterized by little time dispersion, whereas horizontal channels may have long multipath spreads. The extent of the spreading is a strong function of depth and the distance between transmitter and receiver.

High delay and delay variance. The propagation speed in the UW-A channel is five orders of magnitude lower than in the radio channel. This large propagation delay (0.67 s/km) and its variance can reduce the system throughput.

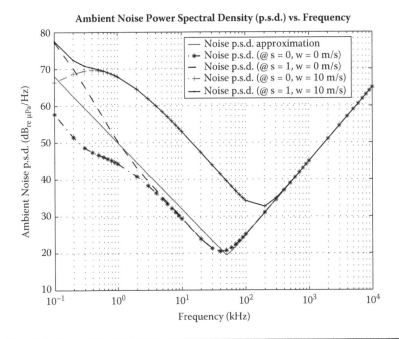

Figure 1.3   Underwater ambient noise power spectrum density $N(f)$ **[dB$_{re\ \mu Pa}$/Hz]** **at different shipping activities** $s = 0, 1$ **and surface wind** $w = 0, 10$ **m/s.**

Specifically, the underwater acoustic propagation speed $q(z, S, t)$ [m/s] is accurately modeled[5] as $q(z, S, t) = 1449.05 + 45.7 \cdot t - 5.21 \cdot t^2 + 0.23 \cdot t^3 + (1.333 - 0.126 \cdot t + 0.009 \cdot t^2) \cdot (S - 35) + 16 \cdot 3.z + 0.18 \cdot z^2$, where $t = T/10$ (T is the temperature in °C), S is the salinity in *ppt*, and z is the depth in km. The above expression provides a useful tool to determine the propagation speed, and thus the propagation delay, in different operating conditions, and yields values in [1460, 1520] m/s, centered around 1500 m/s.

Doppler spread. The Doppler frequency spread can be significant in UW-A channels,[1] causing a degradation in the performance of digital communications: transmissions at a high data rate cause many adjacent symbols to interfere at the receiver. The Doppler spreading generates two effects: a simple frequency translation and a continuous spreading of frequencies, which constitutes a non-shifted signal. While the former is easily compensated at the receiver, the effect of the latter is harder to be compensated for.

Most of the described factors are caused by the chemical-physical properties of the water medium such as temperature, salinity, and density, and by their spatio-temporal variations. These variations cause the acoustic channel to be *highly temporally and spatially variable*. In particular, the horizontal channel is by far more rapidly varying than the vertical channel, especially in shallow water.

1.3 Communication Architectures

In this section, we present some reference communication architectures for underwater acoustic sensor networks, which constitute a basis for discussion of the challenges associated with the underwater environment.

1.3.1 2D Underwater Sensor Networks

A reference architecture for 2D underwater networks is shown in Figure 1.4. A group of sensor nodes are anchored to the bottom of the ocean. Underwater sensor nodes are interconnected to one or more *underwater gateways* (uw-gateways) by means of wireless acoustic links. Uw-gateways are network devices in charge of relaying data from the ocean bottom network to a surface station. To achieve this objective, they are equipped with two acoustic transceivers, namely a *vertical* and a *horizontal* transceiver. The horizontal transceiver is used by the uw-gateway to communicate with the sensor nodes in order to: (i) send commands and configuration data to the sensors (uw-gateway to sensors), and (ii) collect monitored data (sensors to uw-gateway). The vertical link is used by the uw-gateways to relay data to a *surface station*. In deep water applications, vertical transceivers must be

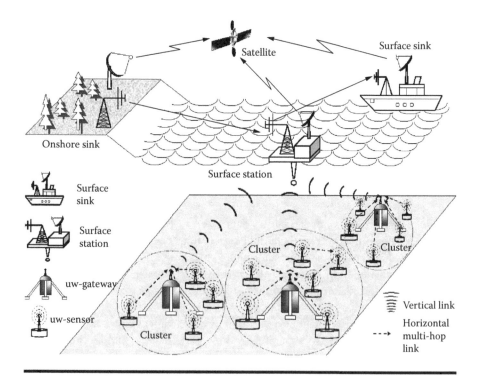

Figure 1.4   2D underwater sensor networks.

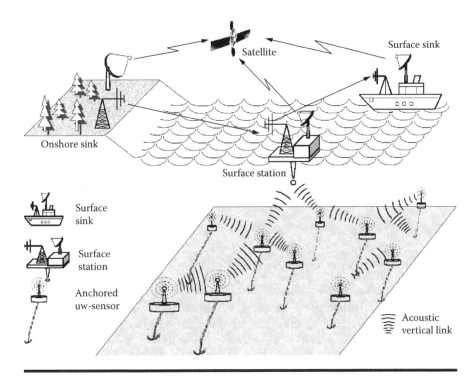

Figure 1.5 3D underwater sensor networks.

long-range transceivers. The surface station is equipped with an acoustic transceiver that is able to handle multiple parallel communications with the deployed uw-gateways. It is also endowed with a long-range RF and/or satellite transmitter to communicate with the *onshore sink* (os-sink) and/or to a *surface sink* (s-sink). In shallow water, bottom-deployed sensors/modems may directly communicate with the surface buoy, with no specialized bottom node (uw-gateway).

1.3.2 3D Underwater Sensor Networks

Three-dimensional underwater networks are used to detect and observe phenomena that cannot be adequately observed by means of ocean bottom sensor nodes, i.e., to perform cooperative sampling of the 3D ocean environment. In 3D underwater networks, sensor nodes float at different depths to observe a phenomenon. In this architecture, given in Figure 1.5, each sensor is anchored to the ocean bottom and equipped with a floating buoy that can be inflated by a pump. The buoy pushes the sensor towards the ocean surface. The depth of the sensor can then be regulated by adjusting the length of the wire that connects the sensor to the anchor, by means of an electronically controlled engine that resides on the sensor.

Sensing and communication coverage in a 3D environment have been rigorously investigated.[8] The diameter and minimum and maximum degrees of the *reachability graph* that describes the network are derived as a function of the communication range, while different degrees of coverage for the 3D environment are characterized as a function of the sensing range.

We presented a statistical analysis for different deployment strategies for 2D and 3D communication architectures for UW-ASNs.[9] Specifically, we determined the minimum number of sensors needed to be deployed to achieve the optimal sensing and communication coverage; we provided guidelines on how to choose the optimal deployment surface area, given a target region; we studied the robustness of the sensor network to node failures, and provided an estimate of the number of redundant sensors to be deployed to compensate for possible failures.

1.3.3 Autonomous Underwater Vehicles

AUVs can function without tethers, cables, or remote control, and therefore they have a multitude of applications in oceanography, environmental monitoring, and underwater resource studies. Previous experimental work has shown the feasibility of relatively inexpensive AUV submarines equipped with multiple underwater sensors that can reach any depth in the ocean. The integration of UW-ASNs with AUVs requires new network coordination algorithms such as

- *Adaptive sampling.* This includes control strategies to command the mobile vehicles to places where their data will be most useful. For example, the density of sensor nodes can be adaptively increased in a given area when a higher sampling rate is needed for a given monitored phenomenon.
- *Self-configuration.* This includes control procedures to automatically detect connectivity holes due to node failures or channel impairment, and request the intervention of an AUV. Furthermore, AUVs can either be used for installation and maintenance of the sensor network infrastructure or to deploy new sensors.

One of the design objectives of AUVs is to make them rely on local intelligence and be less dependent on communications from online shores.[10] In general, control strategies are needed for autonomous coordination, obstacle avoidance, and steering strategies. Solar energy systems allow increasing the lifetime of AUVs, i.e., it is not necessary to recover and recharge the vehicle on a daily basis. Hence, solar-powered AUVs can acquire continuous information for periods of time on the order of months. A reference architecture for 3D UW-ASNs with AUVs is shown in Figure 1.6.

Several types of AUVs exist as experimental platforms for underwater experiments. Some of them resemble small-scale submarines (such as the Odyssey-class AUVs developed at the Massachusetts Institute of Technology [MIT]). Others are

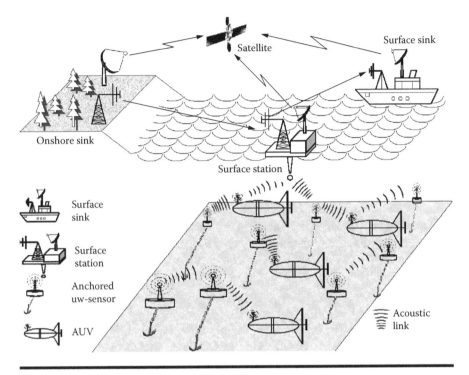

Figure 1.6   3D underwater sensor networks with AUVs.

simpler devices that do not encompass such sophisticated capabilities. For example, *drifters* and *gliders* are oceanographic instruments often used in underwater explorations. Drifter underwater vehicles drift with local current and have the ability to move vertically through the water column, and are used for taking measurements at preset depths.[11] Underwater gliders[12] are battery-powered autonomous underwater vehicles that use hydraulic pumps to vary their volume by a few hundred cubic centimeters in order to generate the buoyancy changes that power their forward gliding.

1.4 Medium Access Control Layer

There has been intensive research on MAC protocols for ad hoc[13] and wireless terrestrial sensor networks[14] in the last decade. However, due to the different nature of the underwater environment and applications, existing terrestrial MAC solutions are unsuitable for this environment. In fact, channel access control in UW-ASNs poses additional challenges due to the peculiarities of the underwater channel, in particular limited bandwidth, very high and variable propagation delays, high bit error rates, temporary losses of connectivity, channel asymmetry, and extensive time-varying multipath and fading phenomena.

Existing MAC solutions are mainly focused on Carrier Sense Multiple Access (CSMA) or Code Division Multiple Access (CDMA). This is because Frequency Division Multiple Access (FDMA) is not suitable for UW-ASN due to the narrow bandwidth in UW-A channels and the vulnerability of limited band systems to fading and multipath. Moreover, Time Division Multiple Access (TDMA) shows limited bandwidth efficiency because of the long time guards required in the UW-A channel. Furthermore, the variable delay makes it very challenging to realize a precise synchronization with a common timing reference.

1.4.1 CSMA-Based MAC Protocols

Slotted FAMA[15] is based on a channel access discipline called Floor Acquisition Multiple Access (FAMA). It combines both carrier sensing (CS) and a dialogue between the source and receiver prior to data transmission. During the initial dialogue, control packets are exchanged between the source node and the intended destination node to avoid multiple transmissions at the same time. Although time slotting eliminates the asynchronous nature of the protocol and the need for excessively long control packets, thus providing savings in energy, guard times should be inserted in the slot duration to account for any system clock drift. In addition, due to the high propagation delay of underwater acoustic channels, the handshaking mechanism may lead to low system throughput, and the carrier sensing may sense the channel idle while a transmission is still going on.

The impact of the large propagation delay on the throughput of selected classical MAC protocols and their variants was analyzed, and the so-called propagation-delay-tolerant collision avoidance protocol (PCAP)[16] was introduced. Its objective is to fix the time spent on setting up links for data frames, and to avoid collisions by scheduling the activity of sensors. Although PCAP offers higher throughput than widely used conventional protocols for wireless networks, it does not provide a flexible solution for applications with heterogeneous requirements.

A distributed energy-efficient MAC protocol tailored for the underwater environment was proposed, whose objective is to save energy based on sleep periods with low duty cycles.[17] The proposed solution is strictly tied to the assumption that nodes follow sleep periods, and is aimed at efficiently organizing the sleep schedules. This protocol tries to minimize the energy consumption and does not consider bandwidth utilization or access delay as objectives.

1.4.2 CDMA-Based MAC Protocols

CDMA is the most promising physical layer and multiple access technique for UW-ASNs. In fact, CDMA is robust to frequency selective fading caused by multipath since it is able to distinguish among signals simultaneously transmitted by multiple devices through codes that spread the user signal over the entire available band. This allows exploiting the time diversity in underwater acoustic channels by leveraging Rake filters[18] at the receiver, so as to compensate for the effect of multipath. In this way, CDMA increases channel reuse and reduces packet retransmissions, which result in decreased battery consumption and increased throughput.

Two code-division spread-spectrum physical layer techniques were compared[19] in shallow water underwater communications, namely Direct Sequence Spread Spectrum (DSSS) and Frequency Hopping Spread Spectrum (FHSS). While in DSSS data is spread using codes with good auto- and cross-correlation properties to minimize the mutual interference, in FHSS different simultaneous communications use different hopping sequences and thus transmit on different frequency bands. Interestingly, it is shown that in the underwater environment FHSS leads to a higher bit error rate than DSSS. Another attractive access technique in the recent underwater literature combines multi-carrier transmission with the DSSS CDMA[20,21] as it may offer higher spectral efficiency than its single-carrier counterpart, and may increase the flexibility to support integrated high data rate applications with different quality of service requirements. The main idea is to spread each data symbol in the frequency domain by transmitting all the chips of a spread symbol at the same time into a large number of narrow subchannels. This way, high data rate can be supported by increasing the duration of each symbol, which reduces ISI. However, multi-carrier transmissions may not be suitable for low-end sensors due to their high complexity.

A MAC solution was also introduced for underwater networks with AUVs.[22] The scheme is based on organizing the network in multiple clusters, each composed of adjacent vehicles. Inside each cluster, TDMA is used with long band guards to overcome the effect of propagation delay. Since vehicles in the same cluster are assumed to be close to one another, the negative effect of very high underwater propagation delay and efficiency loss, which is caused by the long time guards required when TDMA is used underwater,[23] are limited. Interference among different clusters is minimized by assigning different spreading codes to different clusters. The proposed solution assumes a clustered network architecture and proximity among nodes within the same cluster.

We proposed a distributed MAC protocol, called UW-MAC,[24] for UW-ASNs. UW-MAC is a transmitter-based CDMA scheme that incorporates a novel closed-loop distributed algorithm to set the optimal transmit power and code length to minimize the near-far effect. It compensates for the effect of multipath by exploiting the time diversity in the underwater channel, thus achieving high channel reuse and low number of packet retransmissions, which results in decreased battery consumption and increased network throughput. UW-MAC leverages a multi-user detector on resource-rich devices such as surface stations, uw-gateways and AUVs, and a single-user detector on low-end sensors. UW-MAC aims at achieving a threefold objective, i.e., guarantee high network throughput, low access delay, and low energy consumption. It is shown that UW-MAC manages to simultaneously meet the three objectives in deep water communications, which are not severely affected by multipath, while in shallow water communications, which are heavily affected by multipath, UW-MAC dynamically finds the optimal trade-off among high throughput, and low access delay and energy consumption, according to the application requirements. Main features of UW-MAC are (i) it provides a *unique and flexible solution* for different architectures such as *static* 2D deep water and 3D shallow water, and architectures with *mobile* AUVs; (ii) it is fully *distributed*, as code and transmit power are distributively selected by each sender without relying on a centralized entity; (iii) it is intrinsically *secure*, as it uses chaotic codes; (iv) it efficiently *supports multicast transmissions*, as spreading codes are decided at the transmitter side; (v) it is *robust* against inaccurate node position and interference information caused by mobility, traffic unpredictability, and packet loss due to channel impairment. The distributed power and code self-assignment problem to minimize the near-far effect is also formulated, and a low-complexity yet optimal solution is proposed. It is worth noting that UW-MAC is the first protocol that leverages CDMA properties to achieve multiple access to the scarce underwater bandwidth, while existing papers analyzed CDMA only from a physical layer perspective.

1.4.3 Open Research Issues

- In case CDMA is adopted, which we advocate, it is necessary to design access codes with high auto-correlation and low cross-correlation properties to achieve minimum interference among users.
- It is necessary to design low-complexity encoders and decoders to limit the processing power required to implement Forward Error Correction (FEC) functionalities.
- Distributed protocols should be devised to reduce the activity of a device when its battery is depleting without compromising on network operation.

1.5 Network Layer

In recent years there has been a great interest to develop new routing protocols for terrestrial ad hoc[25] and wireless sensor networks.[26] However, there are several drawbacks with respect to the suitability of the existing terrestrial routing solutions for underwater networks. The existing routing protocols are divided into three categories, namely *proactive*, *reactive*, and *geographical* routing protocols.

Proactive protocols (e.g., DSDV,[27] Destination Sequenced Distance Vector; OLSR[28] Optimized Link State Routing) cause a large signaling overhead to establish routes for the first time and each time the network topology is modified because of mobility or node failures, since updated topology information must be propagated to all network devices. This way, each device is able to establish a path to any other node in the network, which may not be needed in UW-ASNs.

Reactive protocols (e.g., AODV,[29] Adhoc On Demand Distance Vector; DSR[30] Dynamic Source Routing) are more appropriate for dynamic environments but incur a higher latency and still require source-initiated flooding of control packets to establish paths. Reactive protocols are unsuitable for UW-ASNs as they also cause a high latency in the establishment of paths, which is further amplified in the underwater by the slow propagation of acoustic signals. Moreover, the topology of UW-ASNs is unlikely to vary dynamically on a short time scale.

Geographical routing protocols (e.g., GFG,[31] Greedy Face Greedy; PTKF[32], Partial Topology Knowledge Forwarding) are very promising for their scalability feature and limited required signaling. However, Global Positioning System (GPS) radio receivers, which may be used in terrestrial systems to accurately estimate the geographical location of sensor nodes, do not properly work in the underwater environment. In fact, GPS uses waves in the 1.5 GHz band that do not propagate in water. Still, underwater devices (sensors, UUVs, UAVs, etc.) need to estimate their current position, irrespective of the chosen routing approach. In fact, it is necessary to associate the sampled data with the 3D position of the device that generates the data, to spatially reconstruct the characteristics of the event. Underwater localization can be achieved by leveraging the low speed of sound in water, which permits accurate timing of signals, and pairwise node distance data can be used to perform 3D localization.[33]

Some recent papers propose network layer protocols specifically tailored for underwater acoustic networks. A routing protocol was proposed that autonomously establishes the underwater network topology, controls network resources, and establishes network flows, which relies on a centralized network manager running on a surface station.[34] The manager establishes efficient data delivery paths in a centralized fashion, which allows the avoiding of congestion and providing some form of quality of service guarantee. Although the idea is promising, the performance evaluation of the proposed mechanisms has not been thoroughly studied.

A routing protocol called vector-based forwarding (VBF)[35] was proposed, which is based on a geographical routing approach and thus does not require state information on the sensors. In VBF, each packet carries the positions of the sender, the destination, and the forwarder. The forwarding path is specified by the so-called *routing vector*, i.e., a vector that connects source and destination. Upon receiving a packet, a node computes its position relative to the forwarder by measuring its distance to the forwarder and the angle of arrival of the signal. Recursively, all the nodes receiving the packet compute their positions. If a node determines that it is close enough to the routing vector (i.e., less than a predefined distance), it includes its own position in the packet and forwards it. Otherwise, it discards the packet. In this way, all packet forwarders form a "routing pipe," and all sensor nodes in the pipe are potential forwarders for the packet. Instead, those nodes that are not close enough to the routing vector, which constitutes the axis of the pipe, do not forward the packet. Packets are thus forwarded along redundant and interleaved paths from source to destination, which makes the protocol robust against packet loss and node failure. The proposed solution can be seen as a form of geographically controlled flooding. However, redundant transmissions are not energy and bandwidth efficient. A localized and distributed self-adaptation algorithm is also proposed to enhance the performance of VBF, which allows the nodes to weigh the benefit of forwarding packets, and accordingly reduce the energy consumption by discarding low benefit packets.

A simple design example of a shallow water network is suggested where routes are established by a central manager based on neighborhood information gathered from all nodes by means of poll packets.[36] However, the routing issues, such as the criteria used to select data paths, are not covered. Moreover, sensors are only deployed linearly along a stretch, while the characteristics of the 3D underwater environment are not investigated.

A long-term monitoring platform for underwater sensor networks consisting of static and mobile nodes was proposed, and hardware and software architectures were described.[37] The nodes communicate point-to-point using a high-speed optical communication system, and broadcast using an acoustic protocol. The mobile nodes can locate and hover above the static nodes for data muling, and can perform useful network maintenance functions such as deployment, relocation, and recovery. However, due to the limitations of optical transmissions, communication is enabled only when the sensors and the mobile mules are in close proximity.

The reliability requirements of long-term critical underwater missions, and the small scale of underwater sensor networks, suggest the devising of routing solutions based on some form of centralized planning of the network topology and data paths, in order to optimally exploit the scarce network resources. For these reasons, we investigated the problem of data gathering for 3D underwater sensor networks at the network layer by considering the interactions between the routing functions and the characteristics of the underwater acoustic channel.[38] We developed a resilient routing solution for long-term monitoring missions, with the objective of guaranteeing survivability of the network to node and link failures. The solution relies on a *virtual circuit* routing technique, where multihop connections are established *a priori* between each source and sink, and each packet associated with a particular connection follows the same path. This requires centralized coordination and leads to a less flexible architecture, but allows exploiting powerful optimization tools on a centralized manager (e.g., the surface station) to achieve optimal performance at the network layer with minimum signaling overhead.

Specifically, the proposed routing solution[38] follows a *two-phase* approach. In the *first phase*, the network manager determines optimal *node-disjoint primary* and *backup* multihop data paths such that the energy consumption of the nodes is minimized. This is needed because, unlike in terrestrial sensor networks where sensors can be redundantly deployed, the underwater environment requires minimizing the number of sensors. Hence, protection is necessary to avoid network connectivity being disrupted by node or link failures. In the *second phase*, an on-line distributed solution guarantees survivability of the network, by locally repairing paths in case of disconnections or failures, or by switching the data traffic on the backup paths in case of severe failures. The emphasis on survivability is motivated by the fact that underwater long-term monitoring missions can be extremely expensive. Hence, it is crucial that the deployed network be highly reliable, so as to avoid failure of missions due to failure of single or multiple devices. The protection scheme proposed can be classified as a dedicated backup scheme with 1:1 path protection, with node-disjoint paths.

We proposed new geographical routing algorithms for the 3D underwater environment,[39] designed to distributively meet the requirements of delay-insensitive and delay-sensitive sensor network applications. The proposed distributed routing solutions are tailored for the characteristics of the underwater environment, e.g., they take explicitly into account the very high propagation delay, which may vary in horizontal and vertical links; the different components of the transmission loss; the impairment of the physical channel; the extremely limited bandwidth; the high bit error rate; and the limited battery energy. In particular, the proposed routing solutions allow achieving two apparently conflicting objectives, i.e., increasing the efficiency of the channel by transmitting a *train* of short packets *back-to-back*, and limiting the packet error rate by keeping the transmitted packets short. The packet-train concept is exploited in the proposed routing algorithms, which allow each node to *jointly* select its best next hop, the transmitted power, and the FEC rate for each packet, with the objective of minimizing the energy consumption, taking the condition of the underwater channel and the application requirements into account.

The first algorithm deals with delay-insensitive applications and tries to exploit links that guarantee a low packet error rate, to maximize the probability that a packet is correctly decoded at the receiver, and thus minimize the number of required packet retransmissions. The second algorithm is designed for delay-sensitive applications. The objective is to minimize the energy consumption, while statistically limiting the end-to-end packet delay and packet error rate by estimating at each hop the time to reach the sink and by leveraging statistical properties of underwater links. In order to meet these application-dependent requirements, each node *jointly* selects its best next hop, the transmitted power, and the forward error correction rate for each packet. Different from the previous delay-insensitive routing solution, next hops are selected by also considering maximum per-packet allowed delay, while unacknowledged packets are not retransmitted to limit the delay. The emphasis on energy consumption is justified by the need for extended lifetime deployments of underwater sensor networks.

There are still several open research issues regarding routing algorithms for underwater networks:

■ For delay-sensitive applications, there is a need to develop algorithms to provide strict latency bounds.
■ For delay-insensitive applications, there is a need to develop mechanisms to handle loss of connectivity without provoking immediate retransmissions. Moreover, algorithms and protocols need to be devised that detect and deal with disconnections due to failures, unforeseen mobility of nodes, or battery depletion.

- Accurate network modeling is needed to better understand the dynamics of data transmission at the network layer. Moreover, realistic simulation models and tools need to be developed.
- Low-complexity acoustic techniques to solve the underwater localization problem with limited energy expenditure in the presence of measurement errors need to be further investigated by the research community.
- Mechanisms are needed to integrate AUVs in underwater networks and to enable communication between sensors and AUVs. In particular, all the information available to sophisticated AUV devices (trajectory, localization) could be exploited to minimize the signaling needed for reconfigurations.

1.6 Transport Layer

A transport layer protocol is needed in UW-ASNs to achieve *reliable transport* of event features, and to perform *flow control* and *congestion control*. Most existing TCP (Transmission Control Protocol) implementations are unsuited for the underwater environment since the flow control functionality is based on a window-based mechanism that relies on an accurate estimate of the Round Trip Time (RTT). The long RTT, which characterizes the underwater environment, would affect the throughput of most TCP implementations. Furthermore, the variability of the underwater RTT would make it hard to effectively set the timeout of the window-based mechanism, which most current TCP implementations rely on.

Existing rate-based transport protocols seem to be unsuited for this challenging environment as well, since they rely on feedback control messages sent back by the destination to dynamically adapt the transmission rate. The long and variable RTT can thus cause instability in the feedback control. For these reasons, it is necessary to devise new strategies to achieve flow control and reliability in UW-ASNs.

A transport layer protocol designed for the underwater environment, Segmented Data Reliable Transport (SDRT),[40] has been recently proposed. SDRT addresses the challenges of underwater sensor networks for reliable data transport, i.e., large propagation delays, low bandwidth, energy efficiency, high error probabilities, and highly dynamic network topologies. The basic idea of SDRT is to use Tornado codes to recover errored packets to reduce retransmissions. The data packets are transmitted block-by-block and each block is forwarded hop-by-hop. SDRT keeps sending packets inside a block before it gets back a positive feedback and thus wastes energy. To reduce such energy consumption, a window control mechanism is adopted. SDRT transmits the packets within the window quickly, and the remaining packets at a lower rate. A mathematical model is developed to estimate the window size and the FEC block size. The performance of SDRT is also illustrated by simulations.

Encoding and decoding using Tornado codes are computation-intensive operations even though Tornado codes use only XOR operations. This leads to increased energy consumption. In SDRT, there is also no mechanism to guarantee the end-to-end reliability as a hop-by-hop transfer mode is used. Each node along the path must first decode the FEC block and then encode it again to transmit it to the next hop. Again, the total computation overhead will be too high for the network. Similarly, for hop-by-hop operations, each sensor must keep calculating the mean values of the window and the FEC block sizes, which can cause a high computational overhead and accordingly higher energy consumption at each sensor. The overhead due to redundant packets will also be high because of high error probabilities. This overhead is dependent on the accuracy in estimating the window size. If the window size is too large, more packets are sent than necessary. In addition, SDRT does not address one of the fundamental challenges for UW-ASN, i.e., shadow zones, and relies on an in-sequence packet-forwarding scheme. While this may be enough for some applications, time-critical data sensors may need to forward packets continuously even in case of holes in the sequence with an out-of-sequence packet delivery mechanism. SDRT is a first attempt to propose a transport protocol for UW-ASN and addresses some of the aforementioned design principles. However, it is still an evolving work and needs further improvements, as it creates redundant transmissions and is computation intensive.

A complete transport layer solution for the underwater environment should be based on the following design principles:

- *Shadow zones.* Although correct handling of shadow zones requires assistance from the routing layer, a transport protocol should consider these cases.
- *Minimum energy consumption.* A transport protocol should be explicitly designed to minimize the energy consumption.
- *Rate-based transmission of packets.* A transport protocol should be based on rate-based transmission of data units as it allows nodes to have flexible control over the rates.
- *Out-of-sequence packet forwarding.* Packets should be continuously forwarded to accelerate the packet delivery process.
- *Timely reaction to local congestion.* A transport protocol should adapt to local conditions immediately, to decrease the response time in case of congestion. Thus, rather than sinks, intermediate nodes should be capable of determining and reacting to local congestion.
- *Cross-layer interaction-based protocol operation.* Losses of connectivity or partial packet losses (i.e., bit or packet errors) should trigger the protocol to take appropriate actions. Therefore, unlike in the layered communications paradigm, transport protocol operations and critical decisions should be supported by the available information from lower layers.
- *Reliability.* A hop-by-hop reliability mechanism surfaces as a prevalent solution as it provides energy-efficient communication. However, there should also be a mechanism to guarantee the end-to-end reliability.

■ *SACK-based loss recovery.* Many feedbacks with acknowledgment (ACK) mechanisms would throttle down the utilization of the bandwidth-limited channel unnecessarily. Thus, the notion of selective acknowledgment (SACK), which helps preserve energy, should be considered for loss scenarios where it is not possible to perform error recovery at lower layers only.

Open research issues for transport layer solutions are given below:

■ New flow control strategies need to be devised to tackle the high delay and delay variance of the control messages sent back by the receivers.
■ New effective mechanisms tailored to the underwater acoustic channel need to be developed to efficiently infer the cause of packet losses.
■ New reliability-metric definitions need to be proposed, based on the event model and on the underwater acoustic channel model.
■ The effects of multiple concurrent events on the reliability and network performance requirements must be studied.
■ It is necessary to statistically model loss of connectivity events to devise mechanisms to enable delay-insensitive applications.
■ It is necessary to devise solutions to handle the effects of losses of connectivity caused by shadow zones.

1.7 Conclusions

In this chapter, we presented an overview of the state of the art in underwater acoustic sensor networks. We described the challenges posed by the peculiarities of the underwater channel with particular reference to monitoring applications for the ocean environment. We discussed characteristics of the underwater channel and outlined future research directions for the development of efficient and reliable underwater acoustic sensor networks. The ultimate objective of this chapter is to bring together researchers from different areas relevant to underwater networks and to encourage research efforts to lay down fundamental bases for the development of new advanced communication techniques for efficient underwater communication and networking for enhanced ocean monitoring and exploration applications.

References

1. M. Stojanovic. Acoustic (Underwater) Communications. In ed. J. G. Proakis, *Encyclopedia of Telecommunications.* John Wiley and Sons, New York, (2003).
2. J. Proakis, J. Rice, E. Sozer, and M. Stojanovic. Shallow Water Acoustic Networks. In ed. J. G. Proakis, *Encyclopedia of Telecommunications.* John Wiley and Sons, New York, (2003).

3. D. Codiga, J. Rice, and P. Baxley. Networked Acoustic Modems for Real-Time Data Delivery from Distributed Subsurface Instruments in the Coastal Ocean: Initial System Development and Performance, *Journal of Atmospheric and Oceanic Technology.* **21**(2), 331–346, (2004).

4. J. Catipovic. Performance Limitations in Underwater Acoustic Telemetry, *IEEE Journal of Oceanic Engineering.* **15**, 205–216, (July, 1990).

5. R. J. Urick, *Principles of Underwater Sound.* (McGraw-Hill, New York, 1983).

6. F. Fisher and V. Simmons. Sound Absorption in Sea Water, *Journal of Acoustical Society of America.* **62**(3), 558–564, (Sept., 1977).

7. R. Jurdak, C. Lopes, and P. Baldi. Battery Lifetime Estimation and Optimization for Underwater Sensor Networks, *IEEE Sensor Network Operations,* (Winter, 2004).

8. V. Ravelomanana. Extremal Properties of Three-dimensional Sensor Networks with Applications, *IEEE Transactions on Mobile Computing.* **3**(3), 246–257, (July/Sept., 2004).

9. D. Pompili, T. Melodia, and I. F. Akyildiz. Deployment Analysis in Underwater Acoustic Wireless Sensor Networks. In *Proc. of ACM International Workshop on Under Water Networks (WUWNet),* Los Angeles, CA, USA, (Sept., 2006).

10. B. Howe, T. McGinnis, and H. Kirkham. Sensor Networks for Cabled Ocean Observatories. In *Geophysical Research Abstracts,* vol. 5, (2003).

11. M. Hinchey. Development of a Small Autonomous Underwater Drifter. In *Proc. of IEEE Newfoundland Electrical and Computer Engineering Conference (NECEC),* Chicago, IL, USA, (Nov., 2004).

12. R. Davis, C. Eriksen, and C. Jones. Autonomous Buoyancy-driven Underwater Gliders. In ed. G. Griffiths, *The Technology and Applications of Autonomous Underwater Vehicles.* Taylor and Francis, London, (2002).

13. S. Kumar, V. S. Raghavanb, and J. Dengc. Medium Access Control Protocols for Ad Hoc Wireless Networks: A Survey, *Ad Hoc Networks (Elsevier).* **4**(3), 326–358, (May, 2006).

14. K. Kredo II and P. Mohapatra. Medium Access Control in Wireless Sensor Networks, *Computer Networks (Elsevier).* **51**(4), 961–994, (Mar., 2007).

15. M. Molins and M. Stojanovic. Slotted FAMA: A MAC Protocol for Underwater Acoustic Networks. In *Proc. of MTS/IEEE Conference and Exhibition for Ocean Engineering, Science and Technology (OCEANS),* Boston, MA, USA, (Sept., 2006).

16. X. Guo, M. R. Frater, and M. J. Ryan. A Propagation-Delay-Tolerant Collision Avoidance Protocol for Underwater Acoustic Sensor Networks. In *Proc. of MTS/IEEE Conference and Exhibition for Ocean Engineering, Science and Technology (OCEANS),* Boston, MA, USA, (Sept., 2006).

17. V. Rodoplu and M. K. Park. An Energy-Efficient MAC Protocol for Underwater Wireless Acoustic Networks. In *Proc. of MTS/IEEE Conference and Exhibition for Ocean Engineering, Science and Technology (OCEANS),* Washington, DC, USA, (Sept., 2005).

18. E. Sozer, J. Proakis, M. Stojanovic, J. Rice, A. Benson, and M. Hatch. Direct Sequence Spread Spectrum Based Modem for Underwater Acoustic Communication and Channel Measurements. In *Proc. of MTS/IEEE Conference and Exhibition for Ocean Engineering, Science and Technology (OCEANS),* Seattle, WA, USA, (Nov., 1999).

19. L. Freitag, M. Stojanovic, S. Singh, and M. Johnson. Analysis of Channel Effects on Direct-Sequence and Frequency-Hopped Spread-Spectrum Acoustic Communication, *IEEE Journal of Oceanic Engineering.* **26**(4), 586–593, (Oct., 2001).

20. D. Kalofonos, M. Stojanovic, and J. Proakis. Performance of Adaptive MC-CDMA Detectors in Rapidly Fading Rayleigh Channels, *IEEE Transactions on Wireless Communications.* **2**(2), 229–239, (Mar., 2003).

21. D. Konstantakos, C. Tsimenidis, A. Adams, and B. Sharif. Comparison of DS-CDMA and MC-CDMA Techniques for Dual-Dispersive Fading Acoustic Communication Networks, *Proc. of IEE Communications.* **152**(6), 1031–1038, (Dec., 2005).
22. F. Salva-Garau and M. Stojanovic. Multi-Cluster Protocol for Ad Hoc Mobile Underwater Acoustic Networks. In *Proc. of MTS/IEEE Conference and Exhibition for Ocean Engineering, Science and Technology (OCEANS)*, San Francisco, CA, USA, (Sept., 2003).
23. I. F. Akyildiz, D. Pompili, and T. Melodia, Underwater Acoustic Sensor Networks: Research Challenges, *Ad Hoc Networks (Elsevier).* **3**(3), 257–279, (May, 2005).
24. D. Pompili, T. Melodia, and I. F. Akyildiz. A Distributed CDMA Medium Access Control for Underwater Acoustic Sensor Networks. In *Proc. of Mediterranean Ad Hoc Networking Workshop (Med-Hoc-Net)*, Corfu, Greece, (June, 2007).
25. M. Abolhasan, T. Wysocki, and E. Dutkiewicz. A Review of Routing Protocols for Mobile Ad Hoc Networks, *Ad Hoc Networks (Elsevier).* **2**, 1–22, (Jan., 2004).
26. K. Akkaya and M. Younis, A Survey on Routing Protocols for Wireless Sensor Networks, *Ad Hoc Networks (Elsevier).* **3**(3), 325–349, (May, 2005).
27. C. Perkins and P. Bhagwat. Highly Dynamic Destination Sequenced Distance Vector Routing (DSDV) for Mobile Computers. In *Proc. of ACM Special Interest Group on Data Communications (SIGCOMM)*, London, UK, (1994).
28. P. Jacquet, P. Muhlethaler, T. Clausen, A. Laouiti, A. Qayyum, and L. Viennot. Optimized Link State Routing Protocol for Ad Hoc Networks. In *Proc. of IEEE Multi Topic Conference (INMIC)*, pp. 62–68, Pakistan, (Dec., 2001).
29. C. Perkins, E. Belding-Royer, and S. Das. Ad Hoc On Demand Distance Vector (AODV) Routing. IETF RFC 3561.
30. D. B. Johnson, D. A. Maltz, and J. Broch. DSR: The Dynamic Source Routing Protocol for Multi-Hop Wireless Ad Hoc Networks. In ed. C. E. Perkins, *Ad Hoc Networking*, pp. 139–172. Addison-Wesley, Reading, MA, (2001).
31. P. Bose, P. Morin, I. Stojmenovic, and J. Urrutia, Routing with Guaranteed Delivery in Ad Hoc Wireless Networks, *ACM-Kluver Wireless Networks (Springer).* **7**(6), 609–616, (Nov., 2001).
32. T. Melodia, D. Pompili, and I. F. Akyildiz, On the Interdependence of Distributed Topology Control and Geographical Routing in Ad Hoc and Sensor Networks, *IEEE Journal of Selected Areas in Communications.* **23**(3), 520–532 (Mar., 2005).
33. V. Chandrasekhar, W. K. Seah, Y. S. Choo, and H. V. Ee. Localization in Underwater Sensor Networks—Survey and Challenges. In *Proc. of ACM International Workshop on UnderWater Networks (WUWNet)*, Los Angeles, CA, USA, (Sept., 2006).
34. G. Xie and J. Gibson. A Network Layer Protocol for UANs to Address Propagation Delay Induced Performance Limitations. In *Proc. of MTS/IEEE Conference and Exhibition for Ocean Engineering, Science and Technology (OCEANS)*, vol. 4, pp. 2087–2094, Honolulu, HI, USA, (Nov., 2001).
35. P. Xie, J.-H. Cui, and L. Lao. VBF: Vector-Based Forwarding Protocol for Underwater Sensor Networks. In *Proc. of Networking*, pp. 1216–1221, Coimbra, Portugal, (May, 2006).
36. E. Sozer, M. Stojanovic, and J. Proakis, Underwater Acoustic Networks, *IEEE Journal of Oceanic Engineering.* **25**(1), 72–83, (Jan., 2000).

37. I. Vasilescu, K. Kotay, D. Rus, M. Dunbabin, and P. Corke. Data Collection, Storage, and Retrieval with an Underwater Sensor Network. In *ACM Conference on Embedded Networked Sensor Systems (SenSys)*, San Diego, CA, USA, (Nov., 2005).

38. D. Pompili, T. Melodia, and I. F. Akyildiz. A Resilient Routing Algorithm for Long-Term Applications in Underwater Sensor Networks. In *Proc. of Mediterranean Ad Hoc Networking Workshop (Med-Hoc-Net)*, Lipari, Italy, (June, 2006).

39. D. Pompili, T. Melodia, and I. F. Akyildiz. Routing Algorithms for Delay-Insensitive and Delay-Sensitive Applications in Underwater Sensor Networks. In *Proc. of ACM Conference on Mobile Computing and Networking (MobiCom)*, Los Angeles, CA, USA, (Sept., 2006).

40. P. Xie and J.-H. Cui. SDRT: A Reliable Data Transport Protocol for Underwater Sensor Networks. University of Connecticut, Technical Report UbiNet-TR06–03, (Feb., 2006).

Chapter 2

Optimal Clustering for Underwater Acoustic Sensor Networks

Liang Zhao, Scott C.-H. Huang,
Qilian Liang, and Xiuzhen Cheng

Contents

In this chapter, we study the optimal clustering in Underwater Acoustic Sensor Networks (UW-ASNs). Although clustering has been well studied for the terrestrial wireless sensor networks (WSN), the unique characteristics of underwater acoustic communications call for a new study. Because the path loss is not only relevant to the distance, but also related to the working frequency, the optimal cluster size for

29

UW-ASNs should be different from the terrestrial WSN, which is determined by the distance between transmission and receiving. We show that the optimal cluster size is mainly determined by the working frequency of the acoustic transmission. In addition, the data aggregation also plays an important role in determining the optimal cluster size. The simulation results agree well with our theoretical analysis.

2.1 Introduction and Motivation

A UW-ASN consists of a variable number of sensors and vehicles that are deployed to perform collaborative monitoring tasks over a given area. To achieve this objective, sensors and vehicles self-organize in an autonomous network that can adapt to the characteristics of the ocean environment.[1-5] Underwater networking is a rather unexplored area, although underwater communications have been experimented with since World War II when, in 1945, an underwater telephone was developed in the United States to communicate with submarines.[2] The main motivation for underwater acoustic (UWA) networks is their relative ease of deployment since they eliminate the need for cables and they do not interfere with shipping activity. Compared to radio waves, sound has superior propagation characteristics in water, making it the preferred technology for underwater communications. In fact, radio waves propagate at long distances through conductive seawater only at extra low frequencies (30–300 Hz), which require large antennae and high transmission power. For example, the Berkeley Mica 2 Motes, the most popular experimental platform in the sensor networking community, have been reported to have a transmission range of 120 cm in underwater at 433 MHz by experiments performed at the Robotic Embedded Systems Laboratory (RESL) at the University of Southern California. Optical waves do not suffer from such high attenuation but are affected by scattering. Moreover, transmission of optical signals requires high precision in pointing the narrow laser beams. Thus, links in underwater networks are based on acoustic wireless communications.[2,8,6,9,13] UW-ASNs are envisioned to enable applications for environmental monitoring of physical indicators and chemical/biological indicators, tactical surveillance applications, disaster prevention, undersea exploration, and assisted navigation, etc. Examples of underwater acoustic networking include the U.S. Navy's experimental Telesonar and Seaweb Program,[12] Distributed Surveillance Sensor Network (DSSN) by SPAWAR System Center in San Diego, and Autonomous Ocean Sampling Network (AOSN) by the Massachusetts Institute of Technology.

Many researchers are currently engaged in developing networking solutions for terrestrial wireless sensor networks. Although there exist many recently developed network protocols for wireless sensor networks, the unique characteristics of the underwater acoustic communication channel, such as limited bandwidth capacity and variable delays make these protocols unsuitable.[11] The underwater channel is impaired by fading, multipath, and Doppler shift. The perturbation of propagation by the variations in sound speed and reflection of the water floor and water surface interfaces results in inhomogeneous insonifaction of the propagation medium and multipath.[7] Multipaths generate delayed parasite echoes and interferences. All these require very efficient and reliable new data communication protocols. In this chapter, we study optimal cluster size in self-organization in UW-ASNs.

Previous studies on underwater communication schemes often use the time-scheduling approach,[10,14] which might be appropriate for smaller networks for its simplicity. However, the flat architecture could also limit the network size. Especially due to the large propagation delay on acoustic link, simple time-scheduling-based schemes are not suitable for larger underwater networks.[15] In[15], Salva-Garau and Stojanovic proposed a clustering scheme for underwater acoustic vehicle networks, which groups the adjacent vehicles into clusters, and uses TDMA (time-division multiple access) within each cluster. Interference among clusters is managed by assigning different spreading codes to adjacent clusters, while scalability is achieved by spatial reuse of the codes. Network operation begins with an initialization phase, and moves on to continuous maintenance during which mobility is managed. They also use simulation analysis to obtain optimal cluster size and transmission power for a network with given density of vehicles. The platform presented by Vasilescu et al.[16] uses both optical and acoustic links for underwater communications. Although optical communications can achieve a higher data rate, its application is limited to short-range point-to-point communications. This platform also makes use of mobile nodes for data muling, which is ideal for high-volume latency-tolerant applications.

Our study is based on random node distribution, but even for those applications where nodes are manually placed, there is room for randomness. Especially in the underwater environment, sensor nodes can easily be moved out of desired places due to water current caused by wind, tide, animal, and human activities, thus, this might leave more room for errors without taking into consideration the deployment randomness.

The remainder of this chapter is organized as follows. Section 2.2 introduces some preliminaries needed for further discussions. We derive the optimum cluster size and analyze its relationship with the working frequency in Section 2.3. Section 2.4 presents the simulations and section 2.5 concludes the chapter.

2.2 Preliminaries

All existing works for terrestrial sensor networks assumed a path loss model inversely proportional to d, where d is the distance between a sender and receiver. This assumption does not capture the specific conditions of UW-ASNs, in which the path loss depends on frequency as well as distance. In underwater acoustics, the transmitted signal pattern has been modeled in various ways, ranging from a cylindrical pattern to a spherical one. Acoustic signals in shallow waters propagate within a cylinder bounded by the water surface and the sea floor, so cylindrical spreading applies for shallow waters. In this section, we provide some preliminaries needed for further discussion.

2.2.1 Underwater Acoustics Fundamentals

Based on the data and formulas of Urick,[17] Jurdak, Lopes, and Baldi[18] derived the following model:

$$SL = TL + 85,$$ (2.1)

where *SL* is the source level and *TL* is the transmission loss. All the quantities in Equation (2.1) are in *dB re μPa*, where the reference value of 1 *μPa* amounts to 0.67×10^{-22} *Watts/cm²*. For cylindrically spread signals, the transmission loss is approximated by[17]

$$TL = 10 \log d + \alpha d \times 10^{-3},$$ (2.2)

where *d* is the distance between source and receiver in meters, and *α* is the frequency dependent medium absorption coefficient. Fisher and Simmons[19] measured the medium absorption in shallow seawater at temperatures between 4°C and 20°C. The average is obtained by[18]

$$\bar{\alpha} = \begin{cases} 0.0601 \times f^{0.8552} & 1 \le f \le 6 \\ 9.7888 \times f^{1.7885} \times 10^{-3} & 7 \le f \le 20 \\ 0.3026 \times f - 3.7933 & 20 \le f \le 35 \\ 0.504 \times f - 11.2 & 35 \le f \le 50. \end{cases}$$ (2.3)

To guarantee the reception quality, the required threshold of *α*, denoted by *ᾶ*, might be chosen larger than *ᾱ*. However, we can generally expect *ᾶ* be a monotonically decreasing function of frequency *f*. In the following, we simply use Equation (2.3) to represent *ᾶ* and in order to emphasize their relationship, *ᾶ* is written as *ᾶ(f)*.

The transmitter power P_t required to achieve an intensity of I_t at a reference distance of 1*m* is expressed as

$$P_t = 2\pi \times 1m \times H \times I_t,$$ (2.4)

where I_t is related to *SL* by

$$I_t = 10^{SL/10} \times 0.67 \times 10^{-18}.$$ (2.5)

Summing up Equations (2.1), (2.2), (2.4), and (2.5), we obtain

$$p_t = CHde^{a(f)d},$$ (2.6)

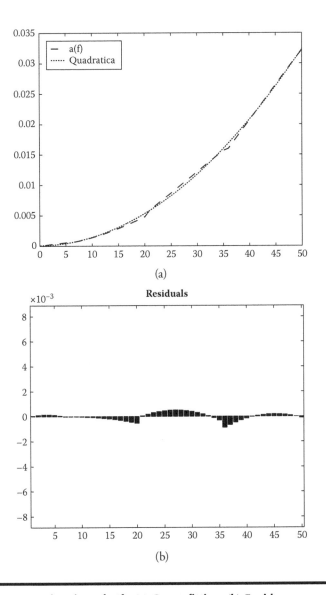

Figure 2.1 **Approximation of a(f). (a) Curve fitting. (b) Residue error.**

$$C \triangleq 2\pi(0.67)10^{-9.5},$$
$$a(\mathrm{f}) \triangleq 0.001\tilde{\alpha}(f)\ln 10, \qquad (2.7)$$

where H is the water depth in meters. We plot $a(f)$ in Figure 2.1(a) in the dashed line. For convenience of further manipulation, we also apply quadratic fitting to

$a(f)$, resulting in

$$a(f) \approx (1.2773e-5)x^2 + (6.5471e-6)x + (0.6802e-5), \qquad (2.8)$$

which is plotted in Figure 2.1(a) in the dotted line. The residuals are plotted in Figure 2.1(b), and the overall RMSE is only $2.9959e-4$ and well within our tolerance of error.

Therefore, to transmit l bits over distance d, the sender's radio expends

$$E_{TX}(l,d) = lE_{elec} + lT_b CHde^{a(f)d} \qquad (2.9)$$

and the receiver's radio expends

$$E_{RX}(l,d) = lE_{elec}, \qquad (2.10)$$

where T_b is the bit duration, and E_{elec} is the unit energy consumed by the electronics to process one bit of message.[20]

2.3 Optimal Clustering

In this section, we analyze the energy consumption in UW-ASNs and then study the relationship between cluster size and energy saving.

2.3.1 Problem Formulation

Clustering has been widely used in pattern recognition, and we use it to obtain the energy-efficient organization for UW-ASNs. Consider a heterogeneous UW-ASN, in which the low-capacity sensors serve as cluster members and are randomly distributed, and the high-capacity sensors serve as cluster heads and are manually positioned. If we determine the optimal cluster size, then the required number of high-capacity sensors and their ideal positions can also be determined. For each bit sent from lower-capacity nodes, the energy consumption at the ith member of the kth cluster for each bit of data is

$$E_{CM(ki)} = E_{elec} + T_b CHr_{ki}e^{a(f)r_{ki}}, \qquad (2.11)$$

where r_{ki} is the distance from the kth cluster head and its ith member. The cluster head collects all the data from its member and then performs data aggregation. On the average, only η bit remains for each incoming bit and η is also referred to as data aggregation ratio. Because the MCUs (Microprogrammed Control Units) used in underwater sensors often work at much lower power than the hydrophones, the energy consumed in data procession is ignored here.[20] Similarly, the energy consumption of the kth high-capacity node is

$$E_{CH(k)} = N_k E_{elec} + N_k \eta (E_{elec} + T_b CH d_k e^{a(f)dk}), \tag{2.12}$$

where N_k is the number of low-capacity nodes in the kth cluster and

$$\sum_{k=1}^{c} N_k = N, \ d_k,$$

where dk is the distance from the kth cluster head to the surface sink.

Considering all c clusters, the overall cost is

$$E_{total} = \sum_{k=1}^{c} \left(E_{CH(k)} + \sum_{i=1}^{N_k} E_{CM(ki)} \right). \tag{2.13}$$

Taking the expected value of the overall energy cost, we obtain $\bar{E}_{total}$ as the objective function.

$$\bar{E}_{total} = k\bar{E}_{CH} + N\bar{E}_{CM}$$

$$= NE_{elec} + NT_b CHE[re^{a(f)r}] + NE_{elec} \tag{2.14}$$

$$+ N\eta \left(E_{elec} + T_b CHE[de^{a(f)d}] \right)$$

Obviously, the determining factor is

$$E[re^{a(f)r}]$$

and

$$E[de^{a(f)d}].$$

Thus, we rewrite Equation (2.14) as

$$\bar{E}_{total} = 2NE_{elec} + NT_b CHJ_{CM}$$

$$+ N\eta (E_{elec} + T_b CHJ_{CH}) \tag{2.15}$$

$$J_{CM} = E[re^{a(f)r}],$$

$$J_{CH} = E[de^{a(f)d}]. \tag{2.16}$$

Suppose the frequency allocation is irrelevant to r, which is the case for most applications in use; $\alpha(f)$ and r are independent. The best cluster size could vary for different deployments. In the following subsection, we will discuss random deployment.

2.3.2 Solution for Random Deployment

Suppose the low-capacity sensors are deployed at random; then their locations would follow the two-dimensional Poisson distribution, i.e., the number of nodes N_A in area A is given by

$$\Pr(N_A) = (\lambda A)^{N_A} e^{-\lambda A} / N_A!, \tag{2.17}$$

where λ is the node density. A useful property of the Poisson process is that if the number of nodes occurring in the area A is N, then the individual outcomes of N nodes are distributed independently and uniformly in the area A. For the single-hop cluster, in which all cluster members can communicate with the cluster head directly, the distance r from a cluster member to the cluster head has the cdf given by

$$F(r) = \frac{\pi r^2}{\pi R_c^2}, \tag{2.18}$$

where R_c is the cluster size. Thus the pdf of r is

$$f(r) = \frac{2r}{R_c^2}. \tag{2.19}$$

$$
\begin{aligned}
J_{CM} &= \int_0^{R_c} r e^{a(f)r} \frac{2r}{R_c^2} dr \\
&= \frac{2r}{R_c^2} \left[\frac{e^{a(f)r}}{a(f)^3} (a(f)^2 r^2 - 2a(f)r + 2) \right]_0^{R_c} \\
&= \frac{2 e^{a(f)R_c}}{a(f)^3} \left(a(f)^2 - \frac{2a(f)}{R_c} + \frac{2}{R_c^2} \right) - \frac{4}{a(f)^3 R_c^2}.
\end{aligned}
\tag{2.20}
$$

Similarly, the cluster heads should also be evenly distributed in the area of interest. Suppose the area of interest is circular with radius R; then the desirable location of cluster heads is depicted by the shadow in Figure 2.2.

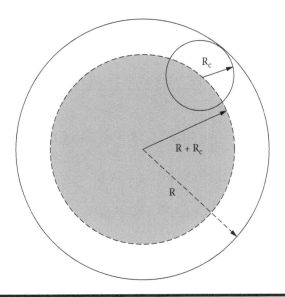

Figure 2.2 Footprint of cluster heads.

$$J_{CH} = \int_0^{R-R_c} r e^{a(f)r} \frac{2r}{R^2} dr$$

$$= \frac{2r}{R^2} \left[\frac{e^{a(f)r}}{a(f)^3} \left(a(f)^2 r^2 - 2a(f)r + 2 \right) \right]_0^{R} {}^{-R_c} \tag{2.21}$$

$$= \frac{2e^{a(f)(R-R_c)}}{a(f)^3 R^2} \left(a(f)^2 (R-R_c)^2 - 2a(f) \right)(R-R_c)$$

$$+ 2 - \frac{4}{a(f)^3 R^2}.$$

By setting the derivative of Equation (2.20) to zero, we obtain

$$\frac{\partial \overline{E}_{total}}{\partial R_c} = \frac{8 + 2e^{aR_c} \left(-4 + 4aR_c - 2a^2 R_c^2 + a^3 R_c^3 \right)}{a^3 R_c^3}$$

$$+ \eta \left(\frac{1}{a^3 R^2} (e^{-aR_c} (-4e^{aR_c} + 2e^{aR} \right. \tag{2.22}$$

$$(a^2 R_c + a(2 - 2aR + aR_c))))$$

$$- \frac{1}{a^2 R^2} (e^{-aR_c} (-4e^{aR_c} + 2e^{aR} (2 + aR(-2 + aR$$

$$+ aR_c (2 - 2aR + aR_c))))) = 0.$$

The solution can be obtained numerically. The second-order derivative of Equation (2.20) is given by

$$
\frac{\partial^2 \overline{E}_{total}}{\partial R_c^2} = \frac{6(-4 + 2e^{aR_c}(2 + aR_c(-2 + aR_c)))}{a^3 R_c^4}
$$

$$
-\frac{1}{a^3 R_c^3}(4(2e^{aR_c}(a^2 R_c + a(-2 + aR_c))
$$

$$
+ 2ae^{aR_c}(2 + aR_c(-2 + aR_c))))
$$

$$
+\frac{1}{a^3 R_c^2}(4a^2 e^{aR_c} + 4ae^{aR_c}(a^2 R_c + a(-2 + aR_c))
$$

$$
+ 2a^2 e^{aR_c}(2 + aR_c(-2 + aR_c)))
$$

$$
+ \eta \left(\frac{e^{-aR_c}\left(4a^2 e^{aR} - 4a^2 e^{aRc}\right)}{a^3 R^2} \right)
$$

$$
-\frac{1}{a^2 R^2}(2e^{-aR_c}(-4ae^{aR_c} + 2e^{aR}(a^2 R_c
$$

$$
+ a(2 - 2aR + aR_c)))) + \frac{1}{aR^2}(e^{-aR_c}(-4e^{aR_c}
$$

$$
+ 2e^{aR}(2 + aR(-2 + aR) + aR_c(2 - 2aR + aR_c)))))).
$$

(2.23)

Substitute Equation (2.22) into Equation (2.23), and it can be shown

$$
\frac{\partial^2 \overline{E}_{total}}{\partial R_c^2} > 0,
$$

(2.24)

which shows $\overline{E}_{total}$ is minimized at the solution of Equation (2.22).

2.4 Simulations

In this section, we compare the validated cluster size using computer simulations. $N = 100$ nodes were uniformly distributed in a circular region with diameter $1000m$ and the water depth was $10m$. The surface sink was set at the center. For a given number of clusters, we used Fuzzy-C-Means (FCM) to form the clusters and then measured the energy consumption of the clustered network. We ran 100

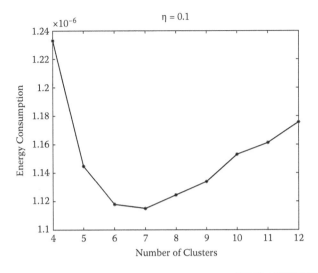

Figure 2.3 E_{total} **versus the number of clusters.**

simulations on randomly generated network topologies and took the average of collected data.

Figures 2.3 to 2.5 show the energy consumption for the given number of clusters with data aggregation ratio η = 0.1, 0.5, and 1.0, respectively. We also solve Equation (2.22) numerically, and translate the cluster radius into the number of

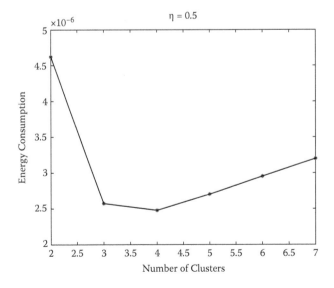

Figure 2.4 E_{total} **versus the number of clusters.**

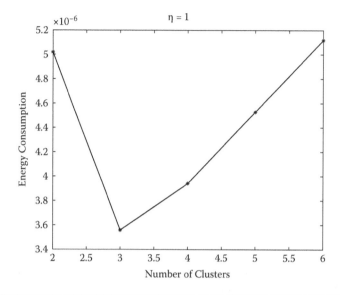

Figure 2.5 E_{total} **versus the number of clusters.**

clusters according to

$$k = \frac{\pi R^2}{\pi R_c^2}.$$ (2.25)

The calculated k's are listed in Table 2.1. Compared to Figures 2.3 to 2.5, Table 2.1 shows that our numerical analysis matches well with the simulation results.

2.5 Conclusions

In this chapter, we studied optimal cluster size in self-organization of UW-ASNs. Although clustering has been well studied for terrestrial WSNs, the unique characteristics of underwater acoustic communications call for a new study. Because the path loss is not only relevant to the distance, but also related to the working frequency, the optimal cluster size for UW-ASNs shows different properties from terrestrial WSNs.

Table 2.1 Calculated Number of Clusters

η	R_c	k
0.1	191.7	6.8
0.5	263.5	3.6
1	293.6	2.9

In addition, data aggregation also plays an important role in determining the optimal cluster size. The simulation results agree well with our numerical analysis.

Acknowledgment

The research of Zhao and Liang was supported in part by the National Science Foundation (NSF) under Grant CNS-0721515. The research of Huang was supported in part by RGC CERG under Grant CityU 113807. The research of Cheng was supported in part by NSF under Grant CNS-0721669.

References

1. F. Akyildiz, D. Pompili, and T. Melodia, "Challenges for efficient communication in underwater acoustic sensor networks," *IEEE ACM Sigbed Review*, vol. 1, no. 2, July 2004.
2. F. Akyildiz, D. Pompili, and T. Melodia, "Underwater acoustic sensor networks: Research challenges," *Ad Hoc Networks (Elsevier)*, vol. 3, no. 3, pp. 257–279, May 2005.
3. P. C. Etter, "Recent Advances in Underwater Acoustic Modelling and Simulation," *Journal of Sound Vibration*, vol. 240, pp. 351–383, Feb. 2001.
4. J. Heidemann, Y. Li, A. Syed, J. Wills, and W. Ye, "Underwater sensor networking: Research challenges and potential applications," USC/ISI, Tech. Rep. ISI-TR-2005-603, 2005.
5. J.-H. Cui, J. Kong, M. Gerla, and S. Zhou, "Challenges: Building scalable mobile underwater wireless sensor networks for aquatic applications," *IEEE Network, Special Issue on Wireless Sensor Networking*, pp. 18, May/June 2006.
6. J. Catipovic, "Performance limitations in underwater acoustic tlemetry," *IEEE Journal of Oceanic Engineering*, vol. 15, pp. 205–216, July 1990.
7. X. Lurton, *An Introduction to Underwater Acoustics: Principles and Applications*, Springer-Praxis Publishing, Chichester, UK, 2002.
8. M. Stojanovic, Acoustic (underwater) communications, in: J. G. Proakis (Ed.), *Encyclopedia of Telecommunications*, Wiley, New York, 2003.
9. P. Xie, J.-H. Cui, and L. Lao, "Vbf: Vector-based forwarding protocol for underwater sensor networks," in *Proceedings of IFIP Networking*, Coimbra, Portugal, May 15–19 2006.
10. D. Makris, *Real-time scheduling and synchronization for the nps autonomous underwater vehicle*, Masters thesis, NAVAL POSTGRADUATE SCHOOL, MONTEREY, CA, 1991.
11. J. G. Proakis, J. A. Rice, and M. Stojanovic, Shallow water acoustic networks. *IEEE Communications Magazine*, (11), 2001.
12. J. A. Rice, "Telesonar signaling and seaweb underwater wireless networks," *Proc. of NATO Symp. New Info. Processing Techniques for Military Systems*, Istanbul, Turkey, Oct. 2000.
13. M. Stojanovic, "Recent advances in high speed underwater acoustic communications," *Oceanic Engineering*, 21(4):12536, 1996.

14. M. Stojanovic, L. Freitag, J. Leonard, and P. Newman, "A network protocol for multiple auv localization," in *Oceans 02 MTS/IEEE*, vol. 1, 2002.
15. F. Salva-Garau and M. Stojanovic, "Multi-cluster protocol for ad hoc mobile underwater acoustic networks," in *OCEANS 2003 Proceedings*, vol. 1, 2003.
16. I. Vasilescu, K. Kotay, D. Rus, M. Dunbabin, and P. Corke, "Data collection, storage, and retrieval with an underwater sensor network," in *SenSys 05: Proceedings of the 3rd International Conference on Embedded Networked Sensor Systems*, New York, NY, USA: ACM Press, 2005, pp. 154–165.
17. R. Urick, *Principles of Underwater Sound*, McGraw-Hill, New York, 1983.
18. R. Jurdak, C. V. Lopes, and P. Baldi, "Battery lifetime estimation and optimization for underwater sensor networks," in *IEEE Sensor Network Operations*, IEEE Press, Winter 2004.
19. F. Fisher and V. Simmons, "Sound absorption in sea water," *Journal of Acoustical Society of America*, vol. 62, p. 558, 1977.
20. "Underwater acoustic modem," available at: www.link-quest.com.

TOPOLOGY CONTROL AND ROUTING

Chapter 3

Topology Control of Three-Dimensional Underwater Wireless Sensor Networks*

S. M. Nazrul Alam and Zygmunt J. Haas

Contents

* This work has been supported in part by the following grants from the National Science Foundation: ANI-0329905 and CNS-0626751, and by the DoD Multidisciplinary University Research Initiative (MURI) program administered by the Air Force Office of Scientific Research under contract number F49620-02-1-0233.

45

Many topology control algorithms in terrestrial wireless sensor networks assume that sensor nodes are deployed on a two-dimensional (2D) plane. This assumption is usually invalid in an underwater sensor network where sensor nodes may be deployed at different depths of the water, thus creating a three-dimensional (3D) network. In this chapter, we investigate the topology control problem for sparse and dense 3D underwater sensor networks. For the sparse network scenario, we assume that nodes are expensive, robust, and can be deployed and maintained in a particular location. In this case, the goal of the topology control algorithm is to deploy the minimum number of sensor nodes such that (a) any point inside the 3D network space is within the sensing range of at least one sensor node, and (b) all the sensor nodes can communicate with each other, possibly over multi-hop paths. For the dense network scenario, we assume that nodes are inexpensive, failure-prone, and their location cannot be maintained for a long time period. In other words, in this case, a large number of redundant nodes are randomly deployed and the topology control algorithm dynamically selects and activates a subset of the nodes to conserve energy while maintaining coverage and connectivity.

3.1 Introduction

Many of the topology control schemes for terrestrial wireless sensor networks assume that sensors are deployed on a 2D plane. Since nodes of a typical terrestrial sensor network are usually deployed in a limited-size area, this assumption is quite often a reasonable approximation. However, in an underwater sensor network, one that is designed to monitor a large 3D space [1], nodes may be deployed at different depths. Topology control in a 3D network is a significantly more challenging problem than in a 2D network. Furthermore, there is no direct and intuitive way to extend some topology control approaches designed for 2D networks to 3D networks, for example, the problem of network deployment with a minimal number of sensor nodes, such that full coverage (i.e., the network can monitor every point of the network) and connectivity (i.e., nodes can communicate with each other) have two very different solutions for 2D and 3D networks. Indeed, the above is a fundamental problem of the sensor network design, because its solution provides a lower bound on the number of nodes needed to achieve full coverage and connectivity. When sensor nodes are expensive, minimizing their number is an important objective, as the total expense of the nodes becomes a major part of the network deployment cost. Another topology control problem arises in a network where sensor nodes are cheap, but failure-prone, and their position cannot be maintained in a particular location. Due to the low cost of the sensor nodes in this case, it is feasible to deploy a large number of those nodes throughout the monitored volume. Therefore, redundancy can be exploited by a topology control algorithm to achieve full coverage and full connectivity. However, to conserve energy, without sacrificing coverage or connectivity, it is not necessary to activate all the sensor nodes at a time. Moreover, by turning off the nodes that are not essential to the network operation at a particular time, the overall network lifetime is extended as well. Consequently, the challenge in this case is to dynamically determine the minimal set of the (still-alive) nodes, so as to maintain full coverage and full connectivity.

Solutions for both of the above problems are known for 2D networks, but these problems have largely been unexplored in the context of 3D networks. In this chapter, we investigate and provide solutions to these two topology control problems for 3D underwater sensor networks.

3.2 Problem Statement

Our common assumptions for both of the problems that we address in this chapter are as follows:

- *Omni-directional, sphere-like sensing model*: The sensing range of an active sensor node is r_s such that an active sensor can reliably monitor a location that is within the distance of r_s from the sensor.
- *Omni-directional, sphere-like communication model*: The communication range of an active node sensor is r_c such that any two active sensor nodes separated by distance of r_c or less can communicate reliably with each other.
- *Homogeneous sensing and communication range*: The parameters r_s and r_c are equal for all the sensor nodes of the network, they have same sensing range, and the communication range of all sensors is also identical.
- *No boundary effects*: The communication and the sensing ranges are much smaller than the length, the width, or the height of the 3D network volume, so that the boundary effects are negligible and hence can be ignored.

■ *Node localization*: The means by which each node is able to estimate its position relative to the other network nodes. This assumption ensures that our solution provides the lower bound on the number of nodes needed to achieve full coverage and connectivity for our topology control problem for sparse networks.* The localization can allow for some small positioning errors.

3.2.1 Topology Control Problem for 3D Sparse Networks

For this case, we assume that a node can be deployed in any position within the volume (or move to that position) as required by the topology control algorithm. The questions that we will answer as part of our topology control study of sparse 3D networks are as follows:

■ Given r_c/r_s, what is the minimum number of sensor nodes, and how should those nodes be deployed in the 3D network, so as to achieve full coverage and full connectivity with all first-tier geographically neighboring nodes?
■ Given r_c/r_s, what is the minimum number of sensor nodes, and how should those nodes be deployed in the 3D network, so as to achieve full coverage and 1-connectivity?[†]

3.2.2 Topology Control Problem for 3D Dense Networks

This topology control problem assumes a large number of redundant nodes, which are densely and uniformly deployed within the to-be-monitored 3D volume. The topology control algorithm partitions the 3D volume into identical cells, keeping one node active in each cell at any time. To do that, the following problems have to be solved:

■ What is the best way to partition a 3D volume into identical virtual cells, so as to minimize the total number of such virtual cells, while the maximum distance between any two points within two neighboring virtual cells does not exceed the communication range r_c?
■ What is the minimum sensing range (relative to the communication range), so that the distance between any two points in a virtual cell does not exceed the sensing range?
■ Design a partitioning scheme so that each sensor node can determine which cell it belongs to in a fast, efficient, and distributed manner.

---

* Of course, the Global Positioning System (GPS) technology does not work under water, so another mechanism is needed for localization. For some options, see [9], [12], [20], and [23].
[†] We say that a network has k-connectivity if every node can communicate with every other node of the network along at least k different node disjoint paths.

3.3 Preliminaries

Our solutions to the above topology control problems are based on the concepts of polyhedron, space-filling polyhedron, Kelvin's conjecture, and Voronoi tessellation. We briefly describe these terms below.

3.3.1 Polyhedron

Any 3D shape consisting of a finite number of polygonal faces* is called a *polyhedron*. The faces of a polyhedron meet in straight-line segments called edges, and the edges meet at points called vertices. Cube, prism, and pyramid are examples of a polyhedron. A polyhedron surrounds a bounded volume in 3D. The two-dimensional analog of a polyhedron is called a *polygon* and the general term for any dimension is *polytope*.

3.3.2 Space-Filling Polyhedron

A *space-filling polyhedron* is a polyhedron that can be used to fill a 3D space without overlaps or gaps (aka space tessellation or tiling). Since our communication and sensing models are sphere-like and 3D spheres do not tessellate, our goal is to find a space-filling polyhedron that best approximates the sphere. In other words, we want to find a space-filling polyhedron such that if each cell is modeled by that polyhedron, then the number of cells required to cover a 3D volume is minimized, where the distance from the center of a cell to its farthest corner (i.e., radius of a cell) is not greater than the sensing range r_s. In many cases, it is not easy to show that a polyhedron has the space-filling property. For example, although Aristotle claimed that the tetrahedron fills space [4], his claim was incorrect [15], and the mistake remained unnoticed until the 16th century [18].

Some of the important results on space-filling polyhedrons are as follows. There are exactly five regular polyhedrons (aka *platonic solids* or *regular solids*) [21]: cube, dodecahedron, icosahedron, octahedron, and tetrahedron, as was proved by Euclid in the last proposition of *The Elements*.† Among them, only cube has the space-filling property [13]. There are only five convex polyhedrons with regular faces having the space-filling property: triangular prism, hexagonal prism, cube, truncated octahedron [21,26], and gyrobifastigium [16]. The rhombic dodecahedron, elongated dodecahedron, and squashed dodecahedron are also space-fillers. A combination of tetrahedrons and octahedrons fills space. In addition, octahedrons, truncated octahedrons, and cubes, combined in the ratio 1:1:3 can also fill space.

---

* A face is part of a plane, i.e., all the points of a face lie on the same plane.
† *The Elements* was written by Euclid about 2300 years ago. The earliest copy, located in Oxford, England, is dated AD 888. *The Elements* contains just theorems and their proofs, without examples, motivations, calculations, or introductions.

However, we impose the restriction that the shape of all the cells should be identical, i.e., only one type of polyhedron is used to fill the space. The motivation for this requirement is twofold:

- ■ Algorithms, especially distributed algorithms, to find the location of nodes are far simpler when one type of polyhedron is used, and
- ■ Since the sensing radius of the polyhedron is fixed, it is unlikely that any significant improvement could be achieved by using two or more types of polyhedrons to fill the space.

3.3.3 Voronoi Tessellation

In three dimensions, for any (topologically) discrete set S of points in Euclidean space, for a given point c ($c \in S$), the set of all points closer to point c than to any other point of S is the interior of a convex polyhedron called the *Voronoi cell* of c. The set of such polyhedrons tessellate the whole space, and is called the *Voronoi tessellation* corresponding to the set S.

3.3.4 Kelvin's Conjecture

In 1887, Lord Kelvin asked the following question [22]: "What is the optimal way to fill a three dimensional space with cells of equal volume, so that the surface area (interface area) is minimized?" This is essentially the problem of finding a space-filling structure having the highest *isoperimetric quotient*. The isoperimetric quotient of a 3D structure with volume V and surface area S is defined as $36\pi V^2/S^3$. Sphere has the largest isoperimetric quotient equal to 1. The answer to Kelvin's question is the 14-sided truncated octahedron with a very slight curvature of the hexagonal faces and with isoperimetric quotient of 0.757. But Kelvin couldn't prove that the structure is optimal. The uncurved truncated octahedron has isoperimetric quotient of 0.753367. For more than a century, Kelvin's solution was generally accepted as correct [27] and has been widely known as *Kelvin's conjecture*. But in 1994, two physicists, Denis Weaire and Robert Phelan, came up with another space-filling structure. It consists of six 14-sided polyhedrons and two 12-sided polyhedrons with irregular faces of equal volume. This structure has 0.3% less surface area than the truncated octahedron [24,25] and its isoperimetric quotient is 0.764. However, currently there is no proof that the structure of Weaire and Phelan is optimal, or that Kelvin's solution is optimal for the case of identical cells.

3.4 Topology Control for Sparse Networks

In this section, we analyze our first topology control problem—that of finding and deploying the minimum number of sensor nodes—so that full coverage and connectivity is achieved.

As shown in [2], this problem can be analyzed from the point of view of the shape of Voronoi cells corresponding to the placement of backbone nodes in 3D space. If all the Voronoi cells are identical and the boundary effects of the covered volume are negligible, then the total number of backbone nodes required is simply the ratio of the volume of the 3D space to be covered to the volume of one Voronoi cell. So minimizing the number of nodes can be achieved if the Voronoi cells have the highest volume such that the radius of its circumsphere does not exceed r_s. Since achieving the highest volume is the goal, the radius of circumsphere must always be equal to r_s, and so the volumes of the circumspheres of all Voronoi cells are the same and equal to $4\pi r_s^3/3$. Finally, the shape of any Voronoi cell in 3D is always a polyhedron. Since we have the restriction that the Voronoi cells have to be identical, all cells have the shape of the same polyhedron. So our problem reduces to the problem of finding the space-filling polyhedron that has the highest ratio of its volume to the volume of its circumsphere. We call this ratio the *volumetric quotient* of the space-filling polyhedron. Since the volume of the circumsphere is the upper bound on the volume of any polyhedron with the same radius, the value of volumetric quotient is always between 0 and 1. Clearly, for a fixed value of r_s, the number of nodes required to cover a 3D space is inversely proportional to the volumetric quotient of the space-filling polyhedron that constitutes a Voronoi cell. So our problem reduces to the problem of finding the space-filling polyhedron that has the highest volumetric quotient. One possible approach is to check all possible space-filling polyhedrons and determine the one with the highest volumetric quotient. However, a rigorous proof that considers all possible space-filling polyhedrons is intractable, as is evident from the fact that Kelvin's problem for the case of a single cell shape remains unproved for more than a century. So, instead, by drawing similarity between our problem and Kelvin's conjecture, we provide some intuition why the truncated octahedron is the most likely solution. Then we choose three other different space-filling polyhedrons that have been used by other researchers in similar problems, and which are reasonable contenders of the truncated octahedron. By comparing these four space-filling polyhedrons, we show that the truncated octahedron has much higher volumetric quotient than the three others polyhedrons.

3.4.1 Similarity with Kelvin's Conjecture

Kelvin's problem for identical cells is essentially finding the space-filling polyhedron that has the highest isoperimetric quotient, while our problem is to find the space-filling polyhedron that has the highest volumetric quotient. Among all 3D shapes, sphere has both the highest isoperimetric quotient and the highest volumetric quotient. We conjecture that for any two space-filling polyhedrons P1 and P2, if P1 has higher isoperimetric quotient than that of P2, then P1 also has higher volumetric quotient than that of P2. Clearly, this is true when we compare any 3D shape with sphere. If this conjecture is true, then the solution to Kelvin's problem for identical cells is essentially the solution to our problem. Since until now, truncated octahedron is the best-known solution to Kelvin's problem for the case of identical cells, we conjecture that the truncated octahedron is, indeed, the most likely solution to our problem too.* Note that in this chapter, we consider the uncurved version of the truncated octahedron, because it is mathematically more tractable than the curved version and the difference between the curved version and the uncurved version is negligible. To further increase the confidence in our solution we compare the truncated octahedron with the other likely space-filling polyhedron contenders.

3.4.2 Seeking Other Polyhedron Contenders

Kepler's conjecture for the sphere packing problem has been proven recently after five centuries of effort, with the FCC lattice being the solution to that problem[†] [14]. The Voronoi tessellation of the FCC lattice is the rhombic dodecahedron. One can attempt to solve our problem using Kepler's problem in the following way. Find the maximal packing of spheres with the Voronoi tessellation corresponding to the centers of the spheres. Define the radius of spheres so that the maximum distance from the center of a sphere to any vertex of the corresponding Voronoi cell is r_s. Phrasing our problem in terms of Kepler's problem suggests that we should choose rhombic dodecahedron as one of the contenders to the truncated octahedron.

The solution to our problem in 2D is the hexagon. The polyhedron that has hexagon as its cross section along all the three axes (i.e., x, y, and z axes) does not have the space-filling property. Two polyhedrons that have the space-filling property with at least one hexagonal cross section are rhombic dodecahedron and hexagonal prism. So we include hexagonal prism in our comparison as well. In fact, two previous works use rhombic dodecahedron [8] and hexagonal prism [11] as the shape of the cell in the context of 3D cellular network. Finally, the most simplistic choice is the cube, and it is the only regular polyhedron that tessellates a 3D space. For notational convenience we term the tessellations and the corresponding node placements of the cube, the hexagonal prism, the rhombic dodecahedron, and the truncated octahedron, as *CB*, *HP*, *RD*, and *TO* models, respectively. In the next subsection, we compare all four models and find that the truncated octahedron has, indeed, higher volumetric quotient than the other choices. Hence the *TO* model requires the fewest number of nodes to cover a particular 3D volume. (We note that [6] provides a proof that implies that the *TO* model actually requires the fewest number of nodes. However, [6] is written in a different context and so does not consider the connectivity issue. As a result, [6] is not directly relevant for solving the problem as a function of r_c/r_s.)

---

[*] We realize that the above argument is not mathematically rigorous.
[†] The problem of finding the densest possible sphere-packing in 3D Euclidean space is known as Kepler's problem.

3.4.3 Analysis

As shown in [2], volumetric quotients of cube, hexagonal prism, rhombic dodecahedron, and truncate octahedron are $2/\sqrt{3}\pi = 0.36755$, $3/2\pi = 0.477$, $3/2\pi = 0.477$, and $24/5\sqrt{5}\pi = 0.68329$, respectively. So *CB*, *HP*, and *RD* models require 85.9%, 43.25%, and 43.25% more nodes than the *TO* model, respectively. Clearly, the number of nodes needed in the *TO* model is significantly smaller than in the other three models. Now, if we consider the connectivity among nodes, then none of the above models works for all values of r_c/r_s. In order to keep any two physically neighboring nodes within the value of r_c, the *CB*, the *HP*, the *RD*, and the *TO* models require that the value of r_c/r_s is at least $2/\sqrt{3} = 1.1547$, $\sqrt{2} = 1.4142$, $\sqrt{2} = 1.4142$, and $4/\sqrt{5} = 1.7889$, respectively. In order to maintain connectivity between any two neighboring nodes, we adjust the models as follows. In the case of the *CB*, the *RD*, and the *TO* models, we set the radius of circumsphere to be

$$R = \min(\sqrt{3}r_c/2, r_s),$$

$$R = \min(r_c/\sqrt{2}, r_s),$$
$$R = \min(r_c\sqrt{5}/4, r_s),$$

respectively. In the case of *HP*, we set each side of the hexagon to

$$a = \min(r_c/\sqrt{3}, r_s\sqrt{2}/\sqrt{3})$$

and the height of each hexagonal prism to

$$b = \min\left(2\sqrt{r_s^2 - a^2}, r_c\right).$$

Then the volume of a cell in all four models for all values of r_c/r_s can be easily determined ([3]) and is shown in Figure 3.1.

Node placement in the *CB*, the *HP*, the *TO*, and the *RD* models can be determined by taking an arbitrary point (x, y, z) as a reference and deploying one node in coordinates

$$\left(x + u\frac{2R}{\sqrt{3}}, y + v\frac{2R}{\sqrt{3}}, z + w\frac{2R}{\sqrt{3}}\right),$$

$$\left(x + \frac{3ua}{2}, y + ua\frac{\sqrt{3}}{2} + va\sqrt{3}, z + wb\right),$$

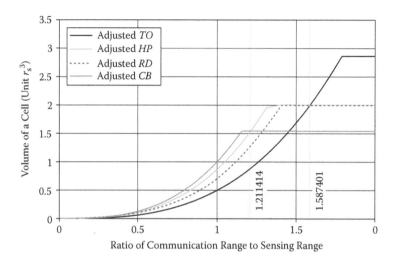

Figure 3.1 Comparison of different node deployment models for different values of r_c/r_s. Volume of a cell in a model is inversely proportional to the number of nodes needed by that model. For $r_c/r_s \geq 1.587401$, we can use the adjusted TO model; for $1.587401 > r_c/r_s \geq 1.211414$, the adjusted HP is the best option, and for $r_c/r_s < 1.211414$, the adjusted CB has the best performance.

$$\left(x + (2u+w)\frac{R}{\sqrt{2}}, y + (2v+w)\frac{R}{\sqrt{2}}, z + wR \right),$$

and

$$\left(x + (2u+w)\frac{2R}{\sqrt{5}}, y + (2v+w)\frac{2R}{\sqrt{5}}, z + w\frac{2R}{\sqrt{5}} \right),$$

respectively, where $u \in \mathbb{Z}$, $v \in \mathbb{Z}$, $w \in \mathbb{Z}$, and $\mathbb{Z}$ is the set of all integers. Clearly, this approach deploys nodes in the entire 3D Euclidean space and creates a network that is infinite along the three axes. In practice, 3D networks are finite. Node placement can be made for a finite network by considering only those coordinates that fall within the coverage space. The same comment applies to other instances in this paper wherever $\mathbb{Z}$ is used.

For $r_c/r_s \geq 4/\sqrt{5}$, all nodes are connected with their physically neighboring nodes and full coverage is always maintained. As a result, there is no overhead to maintain full connectivity in this case. However, when the value of r_c/r_s is smaller, a significant number of extra nodes has to be deployed to guarantee full connectivity, even after full coverage has been already achieved. If we relax the requirement of full connectivity and replace it with 1-connectivity requisite, then communication among distant nodes in general calls for longer routes and failure of nodes could lead to network partition. Consequently, there is a trade-off between network latency and the number of network nodes, or between reliability and the number of nodes.

Relaxing the requirement of full connectivity with all first-tier neighboring nodes is appropriate when the nodes are expensive and reliable, and for a small value of r_c/r_s. In this case, a strip-based node placement strategy can provide full coverage and 1-connectivity, requiring a smaller number of nodes [3].

The strip-based placement strategy is as follows: Deploy nodes as strips such that the distance between any two nodes in a strip is

$$\alpha = \min\{r_c, 4r_s/\sqrt{5}\}$$

and the distance between two parallel strips in a plane is

$$\beta = 2\sqrt{r_s^2 - (\alpha/4)^2}.$$

Set distance between two planes of strips as

$$\beta/2 = \sqrt{r_s^2 - (\alpha/4)^2}$$

and deploy strips such that a strip of one plane is placed between two strips of a neighboring plane (Figures 3.2 and 3.3). The distance between two neighboring nodes that reside in two different planes is

$$\gamma = \sqrt{\beta^2/2 + \alpha^2/4}.$$

This deployment of sensors can be achieved by taking a reference point (x, y, z) and placing a node at each of the coordinates

$$(x + u\alpha + w\gamma\cos\theta, y + v\beta + w\gamma\cos\theta, z + w\gamma\cos\theta),$$

where

$$\theta = \cos^{-1}\left(\sqrt{1/3}\right) \quad \text{and} \quad u \in \mathbb{Z}, v \in \mathbb{Z}, w \in \mathbb{Z}.$$

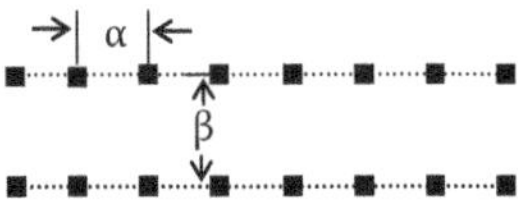

Figure 3.2   Nodes in one plane.

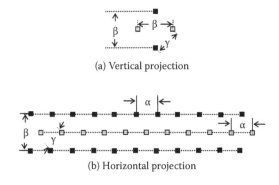

(a) Vertical projection

(b) Horizontal projection

Figure 3.3 Horizontal and vertical projection of nodes in two different planes (nodes with the same shade are from the same plane).

Unless $\beta \leq r_c$ or $\gamma \leq r_c$, this strip-based approach only ensures connectivity among nodes in the same strip. In order to ensure connectivity between strips, additional nodes are needed inbetween the strips. We can achieve 1-connectivity by placing auxiliary nodes such that any two neighboring nodes in two strips are connected (Figure 3.4). However, 2-connectivity can only be achieved by placing auxiliary nodes at the two endpoints of the strips along the boundary of the network. Unless $\beta \leq r_c$ or $\gamma \leq r_c$, there is no way to achieve 3- or higher connectivity without deploying a large number of auxiliary nodes.

For all topology-control algorithms (i.e., node placement strategies) described above, the input to an algorithm is the sensing range r_s, the communication range r_c, and a reference point (x, y, z). These parameters can be hard coded in each sensor node before network deployment. Even if a reference point is not readily available to all nodes, this topology-control algorithm can still be executed in a distributed manner by first selecting a leader (using any standard leader selection algorithm [19]) and then using the location of the leader as the reference point. The leader can broadcast the reference point to all network nodes. Alternatively, a sink can act as a leader. Note that here we assume that the orientation of x, y, z axes is known to all nodes such as from its localization component.

For completeness and comparison, we provide research results for the same problem in the context of 2D networks. Under the similar assumptions (communication and sensing are circle-like in 2D), for $r_c/r_s \geq \sqrt{3}$, placing nodes at the centers of the cells of a regular hexagonal tessellation* of a 2D-plane with radius of r_s, provides both full coverage and connectivity with a minimum number of sensor nodes [5,17,30]. Strip-based deployment (described in [5]) that provides 1-connectivity in 2D networks is useful for $r_c/r_s \geq \sqrt{3}$.

---

* Or, equivalently, at the vertices of a triangular lattice.

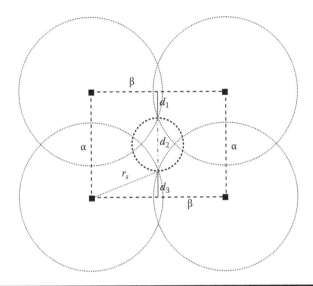

Figure 3.4 **Cross section along a plane. Inner dotted circle is the area covered by a node placed in the planes one level above and below.**

3.5 Topology Control for Dense Networks

In this section, we consider the scenario where sensor nodes' locations cannot be maintained in specific positions, for example, due to ocean current, gravity, or marine animals. The topology control algorithm assumes that the sensor nodes are randomly deployed. Although, due to this random deployment, full coverage and connectivity cannot be guaranteed, full connectivity is more likely if there are more redundant nodes, i.e., denser deployment. Keeping all redundant nodes active increases the consumption of valuable energy and may also increase the network congestion because of redundant message transmission. Consequently, a goal in this type of networks is to dynamically select at any time a subset of the nodes to act as active nodes in a distributed fashion in real time. This approach of putting to sleep a subset of nodes in a sensor network to preserve energy is a well-studied approach in terrestrial sensor networks [7,10,28,29]. One simple way to determine the set of active nodes is to partition the network space into cells, keeping one node active in each cell. In order to make the selection process distributed, we also impose the restriction that cells should be identical. Clearly, the smaller the number of active nodes at a time, the higher the energy savings. However, maintaining full connectivity requires that the maximum distance between the active nodes of any two first-tier neighboring cells cannot exceed the communication range. Since an active node can be located anywhere inside a cell, the maximum distance between any two points of two first-tier neighboring cells must be less than or equal to the transmission radius. One major work in this context is geographic adaptive fidelity (GAF) [28]. Although GAF is proposed to maintain fidelity of routing in a 2D wireless ad hoc network, the concept can easily be extended to a 2D wireless sensor network to maintain coverage and connectivity. GAF divides a 2D network into squared virtual cells (aka grids) and keeps one node active in each cell. It can be shown that GAF performs better when the shape of a virtual cell is a hexagon instead of a square. The energy savings in GAF depends heavily on the choice of the partitioning scheme, because the number of active nodes at a time is equal to the number of total virtual cells. Clearly, the hexagonal partitioning scheme of 2D networks is not applicable in 3D networks. In this section, we investigate and provide a solution for 3D networks.

It should be noted that any criticism of the GAF approach in 2D networks applies to our scheme in 3D networks as well. For example, even the best possible partitioning scheme may require a more than optimal number of active nodes to achieve full coverage and connectivity. However, any scheme that always achieves full coverage and connectivity with the minimum number of active nodes needs a centralized scheduling approach, which is not always practical in a large network. Our scheme treats all nodes in a virtual cell as equivalent from the coverage and the connectivity point of view, and the scheme works well only in a network where nodes are densely and uniformly deployed. In a network where there are cells with no nodes, it is no longer true that any node can be selected as an active node in the other cells [7]. As a result, our solution works only for sufficiently dense networks where at least one live node is present in each cell.

Like the topology-control problem of sparse networks described in the previous section, our focus here is on the four most common polyhedrons that tessellate a 3D space: cube, hexagonal prism, rhombic dodecahedron, and truncated octahedron. However, unlike hierarchical networks, in our case the arrangement of cells is also important as the distance between any two points of two neighboring cells must be within the communication range. For truncated octahedron and rhombic dodecahedron only one arrangement of cells is possible—the regular 3D space tessellation. On the other hand, for cube and hexagonal prism, an alternate arrangement of cells is possible that asymptotically requires fewer nodes than the regular 3D space tessellation. We refer to these alternate arrangements of cube and hexagonal prism as *Alt-CB* and *Alt-HP*. The regular 3D space tessellation of cube, hexagonal prism, rhombic dodecahedron, and truncated octahedron shaped cells are referred to as CB_d, HP_d, RD_d, and TO_d models, respectively (we add subscript d for dense networks to differentiate them from models used for sparse network). Figure 3.5 shows these 3D partitioning schemes graphically.

3.5.1 Analysis

In this section, we derive the maximum radius of a cell and the relationship between the sensing range and the communication range for each model. We also provide an algorithm that allows the sensor nodes to determine their respective cell ID; these IDs are needed for distributed implementation of the topology-control algorithm.

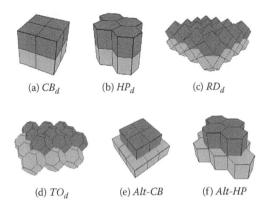

(a) CB_d (b) HP_d (c) RD_d

(d) TO_d (e) *Alt-CB*   (f) *Alt-HP*

Figure 3.5 The 3D partitioning schemes.

Finally, we compare different models in terms of the required number of active nodes at any time and the network lifetime.

3.5.1.1 Maximum Radius of a Cell

Given a fixed transmission radius r_t, the maximum radii of cells in CB_d, Alt-CB, HP_d, Alt-HP, RD_d, and TO_d models are calculated below.

> CB_d *model:* A cell has 26 first-tier neighboring cells—six *Type* 1_{CB} neighboring cells, each sharing one whole side of a cube; 12 *Type* 2_{CB} neighboring cells, each sharing a common line; and eight *Type* 3_{CB} neighboring cells, each sharing just a common point with the cell. Suppose that the radius of a cube is R. Then the largest distance between any point in the cell and any point in a *Type* 1_{CB} neighboring cell is

$$R2\sqrt{2} = 2.828427R.$$

For *Type* 2_{CB} and *Type* 3_{CB} neighbors, it is

$$R2\sqrt{3} = 3.4641R$$

and $4R$, respectively. So an active node of a cell can communicate with active nodes of all first-tier neighboring cells if the maximum radius of a cell in CB_d model is

$$r = \frac{r_c}{\max(2\sqrt{2}, 2\sqrt{3}, 4)} = \frac{r_c}{4} = 0.25r_c.$$

Alt-CB model: A cell has 16 first-tier neighboring cells—four *Type* 1_{Alt-CB} neighboring cells, each sharing one whole side of a cube; four *Type* 2_{Alt-CB} neighboring cells, each sharing a common line; and eight *Type* 3_{Alt-CB} neighboring cells, each sharing one quarter of one side of the cell. The largest distances for *Type* 1_{Alt-CB}, *Type* 2_{Alt-CB}, and *Type* 3_{Alt-CB} cells are

$$R2\sqrt{2},$$

$$R2\sqrt{3},$$

and

$$R\sqrt{34/3},$$

respectively. So the maximum radius of an *Alt-CB* cell is

$$r = \frac{r_c}{\max\left(2\sqrt{2}, 2\sqrt{3}, \sqrt{34/3}\right)} = \frac{r_c}{2\sqrt{3}} = 0.288675 r_c.$$

HP$_d$ model: A cell has 20 first-tier neighboring cells—six *Type* 1_{HP} neighboring cells, each sharing a common square plane; two *Type* 2_{HP} neighboring cells, each sharing a common hexagonal plane; and 12 *Type* 3_{HP} neighboring cells, each sharing a common line with the cell. Suppose that each side of a hexagonal face of an *HP$_d$* cell is of length a, and its height is h. In an *HP$_d$* cell with optimal height,

$$h = a\sqrt{2}.$$

So the radius of an *HP$_d$* cell is

$$R = \sqrt{a^2 + \frac{a^2}{2}} = a\sqrt{\frac{3}{2}}.$$

Thus, the maximum distances from any point in the cell to any point in a *Type* 1_{HP}, *Type* 2_{HP}, and *Type* 3_{HP} neighbor are

$$\sqrt{(a\sqrt{13})^2 + h^2} = R\sqrt{10},$$

$$\sqrt{(2a)^2 + (2h)^2} = R\sqrt{8},$$

and

$$\sqrt{(a\sqrt{13})^2 + (2h)^2} = R\sqrt{14},$$

respectively. As a result, an active node of a cell can communicate with active nodes of all neighboring cells if the maximum radius of a cell in the HP_d model is

$$r = \frac{r_c}{\max(\sqrt{10},\sqrt{8},\sqrt{14})} = \frac{r_c}{\sqrt{14}} = 0.26726r_c.$$

Alt-HP model: A cell has 12 first-tier neighboring cells—six *Type* $1_{Alt\text{-}HP}$ neighboring cells, each sharing a square plane, and six *Type* $2_{Alt\text{-}HP}$ neighboring cells, each sharing one-third of a hexagonal plane with the cell. Maximum distances for *Type* $1_{Alt\text{-}HP}$ and *Type* $2_{Alt\text{-}HP}$ neighbors are

$$\sqrt{(a\sqrt{13})^2 + h^2} = R\sqrt{10}$$

and

$$\sqrt{(3a)^2 + (2h)^2} = R\sqrt{34/3},$$

respectively. So the maximum radius of a cell in *Alt-HP* model is

$$r = \frac{r_c}{\max(\sqrt{10},\sqrt{34/3})} = \frac{r_c}{\sqrt{34/3}} = 0.297r_c.$$

RD_d model: A cell has 18 first-tier neighboring cells—six *Type* 1_{RD} neighboring cells, each sharing just a point, and 12 *Type* 2_{RD} neighboring cells, each sharing a plane with the cell. Maximum distances for *Type* 1_{RD} and *Type* 2_{RD} neighbors are $4R$ and $R\sqrt{10}$, respectively. So the maximum radius of a cell in the RD_d model is

$$r = \frac{r_c}{\max(4,\sqrt{10})} = \frac{r_c}{4} = 0.25r_c.$$

TO_d model: A cell has 14 first-tier neighboring cells—six *Type* 1_{TO} neighboring cells, each sharing a common square plane, and eight *Type* 2_{TO} neighboring

cells, each sharing a common hexagonal plane with the cell. Maximum distances for *Type* 1_{TO} and *Type* 2_{TO} neighbors are

$$\frac{2R}{\sqrt{5}}\sqrt{17}$$

and

$$\frac{2R}{\sqrt{5}}\sqrt{14},$$

respectively. So the active node of a cell can communicate with active nodes of all neighboring cells if the maximum radius of a cell in the TO_d model is

$$r = \frac{r_c}{\max\left(\frac{2\sqrt{17}}{\sqrt{5}}, \frac{2\sqrt{14}}{\sqrt{5}}\right)} = \frac{r_c\sqrt{5}}{2\sqrt{17}} = 0.271163r_c.$$

3.5.1.2 Relationship between Sensing Range and Communication Range

Since an active node can be located anywhere inside a cell, while it must be able to sense any point inside the cell, the sensing range must be at least equal to the maximum distance between any two points of a cell. This maximum distance equals twice the cell's radius. So the minimum sensing ranges of a cell in the CB_d, the *Alt-CB*, the HP_d, the *Alt-HP*, the RD_d, and the TO_d models are

$$2 \times \frac{r_c}{4} = 0.5r_c,$$

$$2 \times \frac{r_c}{2\sqrt{3}} = 0.57735r_c,$$

$$2 \times \frac{r_c}{\sqrt{14}} = 0.53452r_c,$$

$$2 \times \frac{r_c}{\sqrt{34/3}} = 0.5940885r_c,$$

$$2 \times \frac{r_c}{4} = 0.5r_c,$$

and

$$2 \times \frac{r_c \sqrt{5}}{2\sqrt{17}} = 0.542326 r_c,$$

respectively.

3.5.1.3 An Algorithm for a Sensor Node to Determine Its Cell ID

Since the technique is similar for all the above models, we provide calculation only for the TO_d model. Suppose that the information sink (IS), where all data are gathered, resides in the center of a virtual cell and its coordinates (x, y, z) are known. Then for the TO_d model, the center of a virtual cell can be expressed as

$$f(u,v,w) = \left(x + (2u + w)\frac{r_c}{\sqrt{17}}, \; y + (2v + w)\frac{r_c}{\sqrt{17}}, \; z + w\frac{r_c}{\sqrt{17}} \right)$$

and the three integer coordinates (u, v, w) can be used as a unique cell ID. Thus, the ID of the cell which contains the IS is $(0, 0, 0)$. As an example, cell ID $(-1, -1, 2)$ has its center at

$$(x, \; y, \; z + 2r_t/\sqrt{17}).$$

To determine its cell ID, a sensor node first determines its coordinates (x_s, y_s, z_s) using a localization scheme. The transmission radius, r_t, is known to every node, as it has been embedded in the sensors before network deployment. The IS broadcasts its coordinates (x, y, z) to all nodes. One way to determine the cell ID (u_s, v_s, w_s) is a brute force method to check all possible values of (u_s, v_s, w_s) and choose the cell whose center has the minimum Euclidean distance from the node. That is,

$$(u_s, v_s, w_s) = \arg\min_{\substack{u \in \mathbb{Z}, \\ v \in \mathbb{Z}, \\ w \in \mathbb{Z}}} \left(x_s - x - (2u + w)\frac{r_t}{\sqrt{17}} \right)^2 + \left(y_s - y - (2v + w)\frac{r_t}{\sqrt{17}} \right)^2$$

$$+ \left(z_s - z - w\frac{r_t}{\sqrt{17}} \right)^2,$$

where $\mathbb{Z}$ is the set of all integers. However, since the value of a square term is never negative, we can set the value of the square terms to zero. Since these values must be integers, we can get two possible integral values for each variable by taking the ceiling (those variables denoted by the subscript h) and the floor (those variables denoted by the subscript l):

$$u_l = \lfloor (x_s - x - z_s + z)\sqrt{17}/2r_t \rfloor, u_h = \lceil (x_s - x - z_s + z)\sqrt{17}/2r_t \rceil,$$

$$v_l = \lfloor r(y_s - y - z_s + z)\sqrt{17}/2r_t \rfloor, v_h = \lceil (y_s - y - z_s + z)\sqrt{17}/2r_t \rceil,$$

$$w_l = \lfloor (z_s - z)\sqrt{17}/r_t \rfloor, w_h = \lceil (z_s - z)\sqrt{17}/r_t \rceil.$$

Thus we have eight possible values of (u_s, v_s, w_s). Each node calculates its distances from each of the eight centers and chooses the minimum one as its cell ID. That is,

$$(u_s, v_s, w_s) = \arg \min_{\substack{u\in\{u_l,u_h\}, \\ v\in\{v_l,v_h\}, \\ w\in\{w_l,w_h\}}} \sqrt{\left(x_s - x - (2u + w)\frac{r_t}{\sqrt{17}}\right)^2 + \left(y_s - y - (2v + w)\frac{r_t}{\sqrt{17}}\right)^2 + \left(z_s - z - w\frac{r_t}{\sqrt{17}}\right)^2}.$$

Since a cell ID is a straightforward function of the location of a sensor, if a sensor knows the location of another sensor, it can readily calculate the cell ID of that sensor. (Further effort to simplify the cell determination process does not work. For example, instead of calculating the distance from each of the eight centers, taking the nearest integer values for u_s, v_s, w_s leads to an incorrect cell ID in almost one quarter of the cases.)

Once sensors have determined their cell ID, sensors with the same cell ID use any standard leader selection algorithms [19] to choose a node, which becomes the active node of that cell. All nodes with the same cell ID are within the communication range of each other. Thus the mechanism for keeping one node active among all the sensors with the same cell ID is the same as in 2D networks, and we omit further discussion.

3.5.1.4 Number of Active Nodes and Network Lifetime

Ignoring boundary effect, the number of cells in a network is inversely proportional to the volume of the network. Since at any time the number of active nodes in a cell is one, the total number of active nodes in a network is equal to the number of cells in the network. The volumes of a cube, a hexagonal prism, a rhombic dodecahedron, and a truncated octahedron of radius R are $8R^3/3\sqrt{3}$, $2R^3$, $2R^3$, and $32R^3/5\sqrt{5}$, respectively. Using the maximum radius calculated before, we obtain the volumes of cells in the CB_d, the *Alt-CB*, the HP_d, the *Alt-HP*, the RD_d, and the TO_d models as

$$8\left(\frac{r_c}{4}\right)^3 \Big/ 3\sqrt{3} = \frac{r_c^3}{24\sqrt{3}}, 8\left(\frac{r_c}{2\sqrt{3}}\right)^3 \Big/ 3\sqrt{3} = \frac{r_c^3}{27}, 2\left(\frac{r_c}{\sqrt{14}}\right)^3 = \frac{r_c^3}{7\sqrt{14}},$$

$$2\left(\frac{r_c}{\sqrt{34/3}}\right)^3 = \frac{3\sqrt{3}r_c^3}{17\sqrt{34}}, 2\left(\frac{r_c}{4}\right)^3 = \frac{r_c^3}{32},$$

and

$$32\left(\frac{r_c\sqrt{5}}{2\sqrt{17}}\right)^3 \Big/ 5\sqrt{5} = \frac{4}{17\sqrt{17}}r_c^3,$$

respectively. So the numbers of active nodes required by the CB_d, the *Alt-CB*, the HP_d, the *Alt-HP*, and the RD_d models are $96\sqrt{3}/17\sqrt{17}$, $108/17\sqrt{17}$, $28\sqrt{14}/17\sqrt{17}$, $4\sqrt{2}/3\sqrt{3}$, and $128/17\sqrt{17}$ times of that of the TO_d model, respectively.

Now, we use a simplified model to calculate the network lifetime for different space partitioning schemes. Since the transmission radius is the same in all the cases, it can be assumed that a node consumes the same amount of power for transmission in the different cell shapes. If we ignore the power consumption discrepancy due to the difference in the number of packets relayed by a node, then the lifetime of an individual node is roughly the same in all the cases. So the lifetime of a cell is proportional to the number of nodes in a cell. As the sensor nodes are uniformly distributed, the number of nodes in a cell is proportional to the volume of the cell. As a result, in general, the ratio of the network lifetimes for different models is the ratio of volumes of the cells in those models. Consequently, the network lifetime of the CB_d model is

$$\frac{17\sqrt{17}}{96\sqrt{3}} = 42.154\%$$

of that of the TO_d model; for *Alt-CB*, it is

$$\frac{17\sqrt{17}}{108} = 64.9\%$$

of that of the TO_d model; for the HP_d model, it is

$$\frac{17\sqrt{17}}{28\sqrt{14}} = 66.9\%$$

of the TO_d model; for *Alt-HP*, it is

$$\frac{3\sqrt{3}}{4\sqrt{2}} = 91.86\%$$

of the TO_d model; and for the RD_d model, the lifetime is

$$\frac{17\sqrt{17}}{128} = 54.76\%$$

of the lifetime of the TO_d model.

3.6 Discussions

Our assumptions of sphere-like sensing, sphere-like communication (equivalent assumption in 2D is a circle-based model), and equal sensing and communication ranges of the sensor nodes are standard assumptions in most network modeling studies. In practical situations, the network designer can conservatively estimate the sensing range and the communication range (i.e., set sensing range and communication range at some fractional level of their actual values), so as to ensure that the above assumptions remain true.

Because of the finite size of all practical networks, our assumption of no boundary effect is not valid. However, if the physical dimensions of the network are significantly larger than the sensing range of the nodes, then the boundary effects are rather small. If the sensing range is sufficiently small compared to the physical dimensions of the network, it is possible to cover any shaped 3D space with any model without significant overhead. For example, Figure 3.6 shows how a 3D cube-shaped space is covered by a network consisting of 16 × 16 × 16 nodes placed using the TO model.

Finally, the adjustable node position assumption is unlikely to be valid in a practical underwater sensor network. Underwater precise positioning is quite difficult because of lack of highly accurate localization techniques. However, our work does not require an absolute positioning mechanism; rather any relative positioning mechanism where a node knows its position relative to the seed node (i.e., the reference point) is sufficient. Again, in many sensor network applications (e.g., detection, monitoring) it is important to identify positional origination of information. Any underwater sensor network that is deployed for such an application must already have some positioning ability. Using this already available position information allows the topology-control algorithm to avoid the extra overhead. Any positioning mechanism is likely to have some error in determining the position of a node. An underwater sensor network designer can accommodate any potential error by setting the sensing range and communication range of the sensor nodes to a fraction of their actual values, with the fraction depending on the magnitude of the positioning mechanism error.

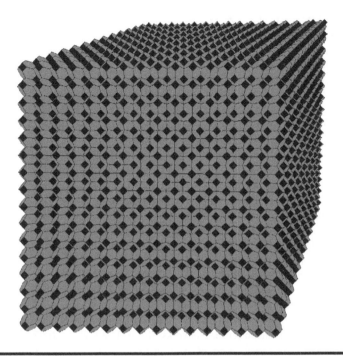

Figure 3.6   An approximation of a cube-shaped space with truncated octahedron-shaped virtual cells.

3.7 Conclusions

In this chapter, we studied topology-control algorithms for the cases of sparse and dense networks. For sparse networks, our topology-control algorithm tries to minimize the number of nodes while maintaining full coverage. For these types of networks, we proposed different algorithms to provide different degrees of communication redundancy. For dense networks, our topology-control algorithm selected a subset of nodes to remain active at any time, so as to ensure coverage and connectivity, while the rest of the live nodes are put to sleep to conserve energy. The selection of active nodes is done in a distributed fashion by partitioning the 3D space into cells and keeping one node active in each cell.

References

1. Akyildiz IF, Pompili D, and Melodia T, Underwater Acoustic Sensor Networks: Research Challenges, *Ad Hoc Networks Journal, (Elsevier)*, March 2005.
2. Alam SMN and Haas Z, "Coverage and Connectivity in Three-Dimensional Networks", In *Proc. of ACM MobiCom*, 2006.

3. Alam SMN and Haas Z, "Coverage and Connectivity in Three-Dimensional Underwater Sensor Networks", Accepted in *Wireless Communication and Mobile Computing (WCMC): Special Issue on Underwater Sensor Networks: Architectures and Protocols*, Wiley, 2008.
4. Aristotle, *On the Heaven*, Vol. 3, Chap. 8, 350 bc.
5. Bai X, Kumar S, Yun Z, Xuan D, and Lai TH, "Deploying Wireless Sensors to Achieve Both Coverage and Connectivity", In *Proc. of ACM MobiHoc*, 2006.
6. Barnes ES and Sloane NJA, "The Optimal Lattice Quantizer in Three Dimensions", *SIAM J. Algebraic Discrete Methods* 4, 30–41, 1983.
7. Basagni S, Carosi A, and Petrioli C, "Sensor DMAC: Dynamic Topology Control for Wireless Sensor Networks," In *Proc. of IEEE VTC*, September 2004.
8. Carle J, Myoupo JF, and Semé D, "A Basis for 3-D Cellular Networks", In *Proc. of the 15th International Conference on Information Networking*, 2001.
9. Chandrasekhar V, Seah WKG, Choo YS, and Ee Hv, "Localization in Underwater Sensor Networks—Survey and Challenges", In *ACM WUWNet*, 2006.
10. Chen B, Jamieson K, Balakrishnan H, and Morris R. "Span: An Energy-Efficient Coordination Algorithm for Topology Maintenance in Ad Hoc Wireless Networks", *Wireless Networks*, 8 (5), 2002.
11. Decayeux C and Semé D, "A New Model for 3-D Cellular Mobile Networks", *ISPDC/HeteroPar*, 2004.
12. Erol M, Vieira LFM, and Gerla M, "Localization with Dive'N'Rise (DNR) Beacons for Underwater Acoustic Sensor Networks," In *ACM WUWNet*, 2007.
13. Gardner M, *The Sixth Book of Mathematical Games from Scientific American*, Chicago, IL: University of Chicago Press, 1984.
14. Hales, TC, "A Proof of the Kepler Conjecture", *Ann. Math.* 162, 1065-1185, 2005
15. Hilbert D and Cohn-Vossen S, *Geometry and the Imagination*. New York: Chelsea, 1999.
16. Johnson NW, *Uniform Polytopes*. Cambridge, England: Cambridge University Press, 2000.
17. Kershner R, "The Number of Circles Covering a Set", *American Journal of Mathematics*, 61:665–671, 1939.
18. Křížek M, "Superconvergence Phenomena on Three-Dimensional Meshes", *International Journal of Numerical Analysis and Modeling*. Vol. 2, No. 1, 2005, pp. 43-56.
19. Lynch N, *Distributed Algorithms*, Morgan Kaufmann Publishers, Wonderland, 1996.
20. Mirza D and Schurgers C, "Energy-Efficient Localization in Networks of Underwater Drifters", In *ACM WUWNet*, 2007.
21. Steinhaus H, *Mathematical Snapshots*, 3rd ed., Oxford: Oxford University Press, 1969.
22. Thomson, W (Lord Kelvin), "On the Division of Space with Minimum Partition Area", *Philosophical Magazine*, 24 (1887) 503–514. http://zapatopi.net/kelvin/papers/on_the_division_of_space.html
23. Tian C, Liu W, Jin J, Wang Y, and Mo Y, "Localization and Synchronization for 3D Underwater Acoustic Sensor Networks", *Ubiquitous Intelligence and Computing*, Lecture Notes in Computer Science series, Springer, Berlin, Germany, 2007.
24. Weaire D, *The Kelvin Problem: Foam Structures of Minimal Surface Area*. London: Taylor and Francis, 1996.
25. Weaire D and Phelan R, "A Counter-Example to Kelvin's Conjecture on Minimal Surfaces", Philosophical Magazine *Letters*, 69, 107–110, 1994.

26. Wells D, *The Penguin Dictionary of Curious and Interesting Geometry*, London: Penguin, 1991.
27. Weyl H, *Symmetry*, Princeton, NJ: Princeton University Press, 1952.
28. Xu Y, Heideman J, and Estrin D, "Geography-Informed Energy Conservation in Ad Hoc Routing", In *Proc. of the 7th ACM MobiCom*, July, 2001.
29. Ye F, Zhong G, Lu S, and Zhang L, "PEAS: A Robust Energy Conserving Protocol for Long-Lived Sensor Networks", *ICDCS '03*, Rhode Island, May 2003.
30. Zhang H and Hou JC, "Maintaining Sensing Coverage and Connectivity in Large Sensor Networks", *Wireless Ad Hoc and Sensor Networks: An International Journal*, Vol. 1, No. 1-2, pp. 89–124, 2005.

Chapter 4

Multipath Virtual Sink Architecture for Underwater Sensor Networks*

Winston K. G. Seah, H. P. Tan, and Pius W. Q. Lee

Contents

---

* This chapter draws material from the authors' publications, which are listed in the references section. Reproduced with permission from IEEE and ACM.

71

Underwater sensor networks promise new opportunities for exploration of the oceans, which cover more than 70% of the earth's surface. Researchers envision the deployment of dense networks of untethered sensors underwater for data acquisition to better understand the underwater environment, while military and security forces see the great potential of using this technology for mine reconnaissance, intrusion detection, and surveillance. Currently, acoustic communications is the only physical layer technique considered viable, and this renders many of the schemes that have been designed for terrestrial sensor networks using radio frequency communications unusable under water. Key challenges include the long propagation delay of the acoustic channel and the severely fluctuating link conditions. The generally adopted single-sink architecture, be it static or mobile, is extremely vulnerable to poor and fluctuating channel conditions, especially when it occurs anywhere en route to the sink. We propose a novel virtual sink approach with multipath data delivery to overcome the adverse link conditions. Multipath tends to lead to contention near the sink, which we avoid with the virtual sink design involving a group of spatially diverse physical sinks. Hence, we are able to exploit the reliability achieved from redundancy provided by multipath data delivery while mitigating the contention. In this chapter, we shall present the proposed architecture and validate its efficacy by studying the performance of relevant data delivery schemes through analysis and simulations. We conclude with a discussion of potential use scenarios and applications that can benefit from this architecture, and future work.

4.1 Introduction

Underwater sensor networks (UWSNs) present many promising opportunities for ocean exploration in search of energy resources, environmental monitoring, and early warning of natural disasters, like tsunamis, but there remain key technical challenges to be overcome. Most notably, the widely used physical layer technology in terrestrial wireless sensor networks, radio frequency (RF) transmission, suffers from severe attenuation in water and has been successfully deployed only at very low frequencies, involving large antenna and high transmission power. Hence, the current viable underwater physical layer technology is acoustic communications.

The salient features of acoustic communications render many schemes that have been designed for RF-based terrestrial sensor networks unusable. Besides having low bandwidth and a propagation delay five orders of magnitude higher than RF in air, the link quality also poses many challenges to underwater communications [1]. A high propagation delay makes it time-consuming to detect packet loss, which has a major impact on the protocols that have been designed for conventional RF-based terrestrial wireless sensor networks where signal propagation delay is negligible. Moreover, the acoustic channel is prone to regional and unpredictable disruptions, resulting in temporal disconnections that can lead to frequent link status updates and excessive re-routing for conventional routing protocols. Such frequent updates drain energy resources and defeat the efforts of link measurement as the time cost of link measurement is significantly higher due to the long propagation delay.

In a sensor network, there are likely to be multiple paths from a sensor to the sink and these paths may or may not be disjointed. It has been shown that routing over multiple disjointed paths increases the packet delivery ratio and achieves timeliness of delivery. However, it comes at a cost of higher energy usage and possibly adding more traffic (packet duplicates) to the network when the link quality is good. Depending on how the paths are selected, there is a strong likelihood of contention occurring among nodes that are on different paths but close to one another. As these paths converge at the sink, the possibility of contention is even higher. Hence, the benefits of multipath routing are diminished when paths are not disjointed and can be totally nullified by the contention among nodes. To reduce the contention due to converging paths, *multiple* sinks [2] can be deployed spatially apart such that all sensors need not send their data to the same sink; rather, they pick the sink that is closest to them to send their data to. Such a multi-sink wireless sensor network architecture opens up new challenges to the data delivery scheme, which is crucial in determining the capacity, energy consumption, and reliability of the network. Using only one of multiple paths can alleviate the interpath contention problem, but an even tougher question of which path to select arises as underwater acoustic links are substantially more unpredictable than RF channels.

We present a multipath virtual sink network architecture using a novel approach of applying fundamental networking concepts. Multipath routing is adopted to provide alternative paths for data to be delivered in order to increase the probability of delivery. To minimize the chances of the multiple paths approaching one another and contending for the shared wireless channel, the paths travel away from one another like a starburst towards multiple sinks deployed along the edges of the sensor network. These sinks collectively form the virtual sink. We validate the design by showing how simple protocols deployed in this architecture can significantly enhance the data delivery in a harsh wireless environment. We validate the efficacy of this architecture by studying the performance of relevant data delivery schemes through analysis and simulations. Lastly, we discuss potential use scenarios and applications that can benefit from this architecture, and future work.

4.2 Challenges in Underwater Networking

In this section, we briefly discuss the challenges faced by underwater sensor networks arising from the physical environment and the use of acoustic communications.

4.2.1 Long Latency and Limited Bandwidth

The underwater acoustic channel is characterized by long latency and low bandwidth. The propagation speed of acoustic waves at 1.5×10^3 m/s is five orders of magnitude slower than RF propagation speeds. Long-range transmission over tens of kilometers can attain bandwidths of only a few kHz while short-range communications in the order of tens of meters may have a bandwidth of a few hundred kHz. In either case, the resultant bit rate is very low, in the order of tens of kbps, at best [3]. Studies conducted on both research as well as commercial modems have shown highly variable link capacities with range* rate product of less than 40 km-kbps [4].

4.2.2 Noise, High Bit Error Rates, and Transmission Loss

The underwater acoustic channel is subjected to various sources of noise. Manmade noise can come from shipping activity and machinery while ambient noise has its origins in hydrodynamics, like wave motion, storms on the surface, etc., and biological sources like seismic activity, fishes swimming, and snapping shrimps. Noise levels can be so high as to cause link blackouts, while intermittent noise results in transmission errors and data loss. Without noise, there is still transmission loss from signal attenuation and geometric spreading; both effects increase with distance, with attenuation also increasing with frequency.

4.2.3 Reliability

The radio propagation speed of 3.0×10^8 m/s makes the RF channel delay negligible as compared to the acoustic channel propagation delay. While automatic repeat request (ARQ) techniques are commonly used in terrestrial wireless communications for packet loss detection, the long propagation delay of the acoustic channel coupled with the low bit rate results in the large bandwidth-delay product problem. Single-hop loss detection will incur at least a round trip delay between two nodes, and error recovery methods like retransmission incur excessive latency and signaling overheads. Multihop scenarios will substantially amplify the effect. When data packets need to be delivered as fast as possible, such mechanisms become a big drawback. It would then appear that forward error correction (FEC) techniques can be applied to provide robustness against errors but at the cost of additional redundant bits competing for the already scarce bandwidth, and the processing needed for encoding and decoding further drains the critical energy resources.

4.2.4 Energy and Cost Constraints

While terrestrial sensor networks are designed with a dense deployment in mind, as they are expected to become very inexpensive, it is not so for UWSNs. The harsh physical underwater environment requires special casing to contain the electronics and often the cost of the casing makes up a larger portion of the total cost of an underwater sensor platform. Deploying a UWSN is also a difficult and costly operation. UWSN nodes are therefore fewer and distances between nodes longer. Coupled with the poor channel conditions, this translates to higher energy usage for longer range transmission and the use of more complex signal processing schemes for reliability. Once deployed, it is often difficult if not impossible to replace the energy source (batteries) after it have been expended. Furthermore, alternative energy sources, like solar energy, are not available in the dark depths of the ocean. The energy constraints make a multihop approach attractive as the energy cost for wireless communication is exponential to the distance between sender and receiver [5]. Given an arbitrary distance, relaying packets using a multihop approach can save significant energy compared to a single long-range transmission. However, it comes with other costs that must be considered too.

4.2.5 Volatile Link Quality

The underwater link quality is extremely volatile, and suffers frequent temporal disconnections due to numerous reasons, such as noise (both manmade and ambient), temperature fluctuations, and severe multipath fading. Current approaches use some form of transmission to measure the link quality, which is time consuming with regard to the huge propagation delay; thus, link quality measurement incurs a considerably high cost or overhead for the underwater scenario. Moreover, a link that has been determined to be of good quality may experience poor link conditions moments later, rendering the costly process useless. The volatile link quality leads to quickly outdated neighborhood status and connectivity information, which is the cornerstone of most routing protocols. Hence, maintaining valid and useful neighborhood status information in the presence of volatile links can be prohibitively costly.

4.3 Network Architecture

The network topology is crucial in determining the network capacity, energy consumption, and, more importantly, the reliability of the network. There must be sufficient robustness and redundancy built into the network to ensure that it continues to function even when a significant portion of the network is temporarily non-operational. When the link quality is poor, the probability of a successful transmission drops exponentially as the number of hops increases. All these issues are taken into account in the design of the multipath virtual sink architecture [6, 7]. Hence, we consider a network topology that is made up of sensing nodes, with local aggregation points distributed among them. These aggregation points will collectively form a wireless mesh relay network that connects to local sinks (as shown in Figure 4.1). Although we have shown a two-tier topology,

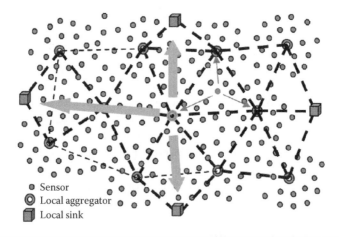

○ Sensor
◉ Local aggregator
▯ Local sink

Figure 4.1 Multipath virtual sink architecture. (From Pius W.Q. Lee and Winston K.G. Seah. A Comparison of Two Data Delivery Schemes for Underwater Sensor Networks. *Proceedings of IEEE OCEANS,* **June 18–21, 2007, Aberdeen, Scotland. © 2007 IEEE. With permission.)**

the number of tiers is flexible and can be dynamically adapted to meet deployment requirements and suit environmental conditions.

4.3.1 Virtual Sink

Our architecture is designed to exploit the advantages of having local aggregation points while avoiding adverse traffic congestion and contention at these nodes. It is assumed that the local sinks are connected via high-speed links (e.g., broadband communications link, or wired high-speed optical fiber) to a network where the resources are more than sufficient to support the communication needs of the various applications. Therefore, the ultimate goal of the wireless sensor network is to ensure that data is delivered to one or more of these local sinks, which collectively forms a *virtual sink*. As the sensing coverage is very dependent on the application's needs and the technologies that are used to develop the sensors, we focus only on the communications aspects of sensors such as the range and bandwidth of the communication link.

4.3.2 Multipath Data Delivery

A robust multipath data delivery scheme provides end-to-end connectivity to the local aggregation points. The scheme aims to maintain n routes to the neighboring local aggregation points and provide local data caching; the value of n adapts to the channel conditions and also the criticality of the data carried in the packet. If the channel is intermittent and bandwidth very limited, it may be better for the nodes to cache data and transmit when the channel conditions are favorable rather than attempt multiple retransmissions. For time-critical data, instead of caching, the scheme will attempt to deliver data over more routes (larger n value) to increase the probability of successful delivery.

Similarly, the local aggregation points form a wireless mesh network that provides multiple paths to multiple local sinks, which collectively form the virtual sink. Congestion at aggregation points (mesh nodes) can occur with simultaneous arrival of a high volume of traffic from sensor nodes, e.g., sensor data arising from the detection of the engine noise of a moving boat on the surface may generate a consecutive burst of sensor traffic arriving at neighboring aggregation points. As the name implies, in-network data aggregation is necessary to handle the congestion at the aggregation points. Likewise, the deployment of redundant nodes (as backup aggregation points) to increase the availability of multiple disjoint paths such that backup routes are readily available can be done, where necessary. This is crucial for sending time-critical delay-intolerant data that cannot be cached until the channel conditions improve. The multipath routing

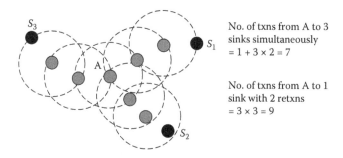

No. of txns from A to 3
sinks simultaneously
= 1 + 3 × 2 = 7

No. of txns from A to 1
sink with 2 retxns
= 3 × 3 = 9

Figure 4.2   Redefining multipath. (From Pius W.Q. Lee and Winston K.G. Seah. A Comparison of Two Data Delivery Schemes for Underwater Sensor Networks. *Proceedings of IEEE OCEANS,* June 18–21, 2007, Aberdeen, Scotland. © 2007 IEEE. With permission.)

protocol will select the appropriate routes from those available to achieve the required service levels.

4.3.3 Redefining Multipath and Re-Transmission

Typical multipath routing protocols set up multiple routes between a pair of communicating nodes [8]. Depending on how the routes are selected, there is a strong likelihood of contention occurring among nodes that are on different routes but close to one another. The possibility of contention is even higher as routes converge at the destination node, nullifying the improvements from the redundancy provided by multipath, and this can be made worse by re-transmissions. As shown in Figure 4.2, we propose that a node (e.g., A) sends a packet simultaneously over spatially diverse routes to multiple sinks (S1, S2, and S3), which form the virtual sink, and as long as a copy of the packet reaches one of these sinks, delivery is successful.

This can be considered as re-transmitting a packet simultaneously instead of sequentially, achieving lower latency and fewer packet transmissions, thus saving energy. The use of spatially diverse paths also reduces the possibility of contention.

4.4 Analysis of Data Delivery over Spatially Diverse Routes [7]

For the proposed multipath virtual sink architecture, we consider the delivery of a single data packet from each local aggregator to local sinks and analyze the performance of multihop routing in terms of latency, transmission reliability, and energy

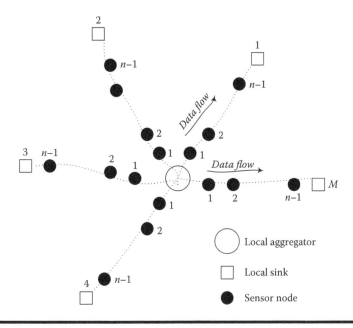

Figure 4.3 Virtual sink model.

consumption, which are important metrics for data dissemination in wireless sensor networks.

Let us consider a source node (local aggregator) for which M spatially diverse paths to M local sinks are available, where each path i comprises n equally spaced hops. We assume that M is small, so that the paths are sufficiently diverse spatially, and therefore the transmissions along each path do not interfere with those of other paths. We associate a parameter $p_i(j)$ for each path i to denote the probability of transmission failure over the *jth* hop, and assume that $p_i(j)$, $1 \le j \le n$, are independent. This model is illustrated in Figure 4.3.

Let t denote the total number of transmission attempts. Assuming all hops to be equidistant with a propagation delay of τ_p and the transmission power is fixed, the energy consumed for the transmission of each packet is proportional to τ. If τ_x denotes the time required for a packet to be delivered per hop, then $\tau_x = \tau_t + \tau_p$, where τ_t is the transmission time required for each packet.

4.4.1 Spatially Diverse Multipath (MP) Routing

With this approach, the local aggregator sends the packet simultaneously over all M paths. Along each path i, the packet reaches the corresponding local sink only if transmissions over all n hops are successful. As long as a copy of the packet reaches one of the local sinks, delivery is successful.

Let us consider the transmission of a single packet over path i. If t_i denotes the total number of transmissions after the first hop, where $0 \le t_i \le n - 1$, then we have the following pmf:

$$P(t_i = t) = \begin{cases} p_i(t+1) \displaystyle\prod_{j=1}^{t} (1 - p_i(j)), & t < n - 1; \\ \displaystyle\prod_{j=1}^{t} (1 - p_i(j)), & t = n - 1. \end{cases}$$

If t_{MP} denotes the *expected* total number of transmissions for a single packet, then we have:

$$t_{MP} = 1 + \sum_{i=1}^{M} \sum_{t=0}^{n-1} t P(t_i = t), \qquad (4.1)$$

where the first term corresponds to the transmission that occurs at the local aggregator, and the second term denotes the subsequent transmissions along each path.

Along any path i, the packet will be successfully received only if transmissions over *all* n hops are successful, and this occurs with the following probability:

$$P_i = \prod_{j=1}^{n} (1 - p_i(j)),$$

since $p_i(j)$, $1 \le j \le n$, are independent.

Hence, the probability that none of the M copies arrives at the local sinks is

$$\prod_{i=1}^{M} (1 - P_i),$$

since the channel behavior over different paths are assumed to be independent. Therefore, the probability of successful packet delivery (which measures the reliability of the routing mechanism) is given by:

$$P_{MP} = 1 - \prod_{i=1}^{M} (1 - P_i). \qquad (4.2)$$

Conditioned on successful delivery, the total packet delay (or latency), T_{MP}, is given as follows:

$$T_{MP} = n\tau_x, \qquad (4.3)$$

since each packet must traverse n hops before arriving at the local sink.

4.4.2 Single-Path (SP) Routing with Stop-and-Wait ARQ

With this approach, the aggregator selects *one* of the *M* paths (e.g., path *i*) to send the packet to the corresponding sink. When a transmission over any hop along the path fails, a simple Stop-and-Wait ARQ strategy is used for re-transmission over that hop.

Consider the packet transmission from node A to node B over a single hop. When node B receives the packet from node A, it sends an acknowledgment (ACK) packet to node A, and proceeds to forward the packet to the node along the next hop. If node A does not receive an ACK packet from node B after a time-out interval, τ_o, it assumes that packet transmission has failed, and initiates a re-transmission.

Let us assume that there are $r_i(j)$ transmission failures (or re-transmissions) over hop j, $r_i(j) \geq 0$, such that

$$r_i = \sum_{j=1}^{n} r_i(j)$$

is the total number of re-transmissions. In this case, the total number of transmissions, $t_{SP,\tau i}$ (including data and ACK packets) is given by:

$$t_{SP,r_i} = 2n + r_i. \tag{4.4}$$

Then, we have the following:

$$P(t_{SP,r_i} = 2n + r_i) = \sum_{\sum_{j=1}^{n} r_i(j) = r_i, r_i(j) \geq 0} \prod_{j=1}^{n} p_i(j)^{r_i(j)} (1 - p_i(j)).$$

For the special case where $p_i(j) = p_i$, the pmf of $t_{SP,ri}$ reduces to a *negative binomial* distribution, given as follows:

$$P(t_{SP,r_i} = 2n + r_i) = \binom{n - 1 + r_i}{n - 1} p_i^{r_i} (1 - p_i)^n.$$

Assuming that, due to energy constraints, a maximum of *R* re-transmissions are permitted, the probability of successful packet delivery is given as follows:

$$P_{SP,R} = \sum_{r_i=0}^{R} P(t_{SP,\tau_i} = 2n + r_i).$$

If r_i re-transmissions occur before the packet is successfully delivered, then the total packet delay is $T_{SP,ri} = r_i \tau_o + n \tau_x$. Hence, conditioned on successful delivery within R re-transmissions, we have the following:

$$P(T_{SP,r_i} = r_i \tau_o + n \tau_x) = \frac{P(t_{SP} = 2n + r_i)}{P_{SP,R}} \tag{4.5}$$

Therefore, the conditional *expected* packet delay, $T_{SP,R}$, is given as follows:

$$
\begin{aligned}
T_{SP,R} &= \sum_{r_i=0}^{R} (r_i \tau_o + n \tau_x) P\left(T_{SP,r_i=r_i\tau_o+n\tau_x}\right) \\
&= n \tau_x + \tau_0 \sum_{r_i=0}^{R} \frac{r_i P(t_{SP} = 2n + r_i)}{P_{SP,R}} \\
&= T_{MP} + \tau_o \sum_{r_i=0}^{R} \frac{r_i P(t_{SP} = 2n + r_i)}{P_{SP,R}}.
\end{aligned}
$$

Hence, conditioned on packet delivery being successful, the latency introduced by single-path routing with ARQ is always *larger* than the multipath routing protocol as long as re-transmissions are permitted.

4.4.3 Numerical Results

In this section, we compare the performance of both routing mechanisms in terms of numerical results.

4.4.3.1 Energy Constraint

We consider the scenario where the bottleneck is the energy constraint of each node. For this scenario, we evaluate the reliability and latency, assuming that both routing mechanisms have *equal* energy consumption. This is done by setting the maximum allowable number of re-transmissions, R, such that $t_{SP,R} = t_{MP}$, i.e., we have the following:

$$2n + R = 1 + \sum_{i=1}^{M} \sum_{t=0}^{n-1} t P(t_i = t). \tag{4.6}$$

We begin with the simplest case where the channel is *spatially invariant*, and is characterized by a single parameter, p, i.e., $p_i(j) = p \ \forall \ i,j$. We plot P_{MP} and $P_{SP,R}$ [with R computed using Equation (4.6)] as a function of p for $M = 4$, $n = \{5,10,15\}$ and $n = 10$, $M = \{4,5,6\}$ in Figure 4.4. We also plot the corresponding degradation

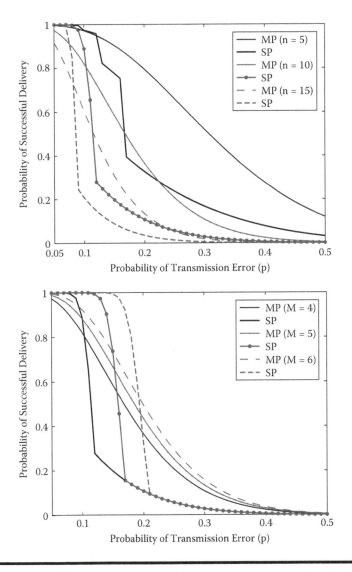

Figure 4.4 Probability of successful packet delivery vs. channel quality for each routing mechanism for *M* = 4, *n* = {5,10,15} (left) and *n* = 10, *M* = {4,5,6} (right).

in latency, conditioned on successful packet delivery, as a factor of τ_o, of the single-path algorithm in Figure 4.5.

For a given value of *M* and *n*, the SP protocol is more reliable (at the expense of high latency) than the MP protocol when the channel is very good. However, there exists a threshold, $p_{thres,r}$ such that for $p > p_{thres,r}$, the MP protocol becomes more reliable while maintaining lower or the same latency compared to the SP protocol. In terms of latency, there exists a threshold, $p_{thres,l}$ such that when $p < p_{thres,l}$, the

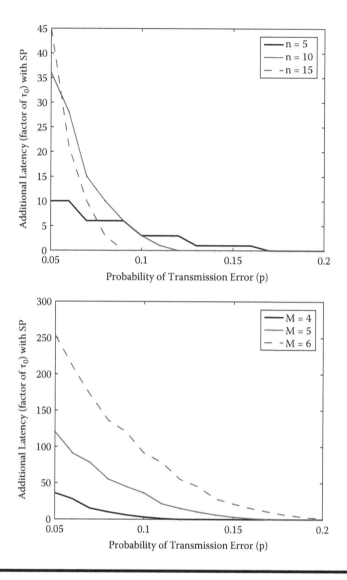

Figure 4.5 Degradation of latency (factor of τ_o) of single-path algorithm vs. channel quality for $M = 4$, $n = \{5,10,15\}$ (left) and $n = 10$, $M = \{4,5,6\}$ (right).

SP protocol always incurs additional latency compared to the MP protocol; however, for $p > p_{thres,l}$, both protocols incur the same latency.

For a fixed M, as n increases, we observe that both $p_{thres,r}$ and $p_{thres,l}$ are reduced. This implies that a larger region (in terms of p) of performance gain achieved by the MP protocol in terms of reliability is traded off with a smaller region of performance gain in terms of latency as the hopcount is increased.

On the other hand, for a given n, as M increases, we observe that both $p_{thres,r}$ and $p_{thres,l}$ are increased. This implies that a smaller region of performance again achieved by the MP protocol in terms of reliability is traded off with a larger region of performance gain in terms of latency as the number of routing paths is increased.

Next, we introduce *spatial-variance* in the channel over the M paths, while maintaining the invariance over the n hops of each path, i.e., $p_i(j) = p_i \ \forall \ i,j, p_i \neq p_k, i \neq$ k. We investigate the impact of this spatial-variance on the reliability and latency of each protocol, *which maintaining* the same level of energy consumption. For $M = 5$, $n = 10$, and $\bar{p} = \{0.1, 0.11, ..., 0.2\}$, we compare P_{MP} and $P_{SP,R}$ as a function of the channel for two cases: (i) spatially invariant channel with $p_i = \bar{p}$ and (ii) spatially variant channel with $p_i = \bar{p} + 0.01(i - 3)$, and the results are shown in the left-hand side (LHS) of Figure 4.6. The corresponding results comparing the latency are shown in the right-hand side (RHS) of Figure 4.6. We note that the *average* probability of transmission failure over all paths in both cases are the same and given by $\bar{p}$.

We observe that the spatial-variance in the channel improves the reliability of each protocol, where the improvement is more significant for the multipath protocol. However, the improvement in reliability for the single-path protocol is achieved at the expense of increased latency.

4.4.3.2 Reliability Constraint

Next, we consider the scenario where the data packet is loss sensitive, and hence, reliability becomes the most important criteria. For this scenario, we evaluate the energy consumption and latency, assuming that both routing mechanisms have *equal* reliability. This is done by setting the maximum allowable number of re-transmissions, R, such that $|P_{MP} - P_{SP,R}|$ is minimized.

We consider the case where the channel is *spatially invariant*, and is characterized by a single parameter, p, i.e., pi(j) = $p \ \forall \ i,j$. We plot t_{MP} and t_{SP} as a function of p for $M = 4$, $n = \{5, 10, 15\}$ and $n = 10$, $M = \{3, 4, 5\}$ in Figure 4.7. We also plot the corresponding degradation in latency, conditioned on successful packet delivery, as a factor of τ_o, of the single-path algorithm in Figure 4.8.

For a given value of M and n, the SP protocol is more energy-efficient than the MP protocol when the channel is very good. However, there exists a threshold, $p_{thres,r}$, such that for $p > p_{thres,r}$, the MP protocol becomes more energy efficient. In terms of latency, the SP protocol always incurs additional latency compared to the MP protocol.

For a fixed M, as n increases, we observe that $p_{thres,r}$ is reduced. This implies that a larger region (in terms of p) of performance gain is achieved

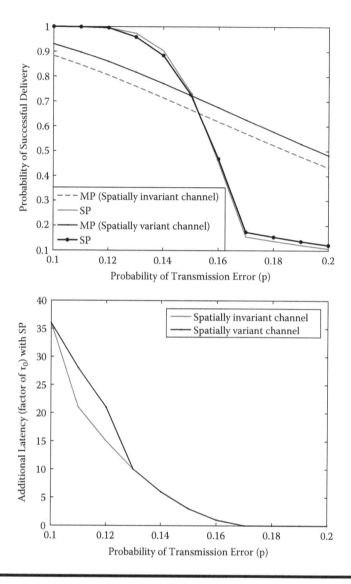

Figure 4.6 Impact of spatial-variance of channel on the probability of successful packet delivery (left) and degradation of latency (factor of τ_o) of SP algorithm (right) vs. channel quality for $M = 5$, $n = 10$.

by the MP protocol in terms of energy efficiency as the hopcount is increased. On the other hand, for a given n, as M increases, we observe that $p_{thres,r}$ is increased. This implies that a smaller region of performance again is achieved by the MP protocol in terms of energy efficiency as the number of routing paths is increased.

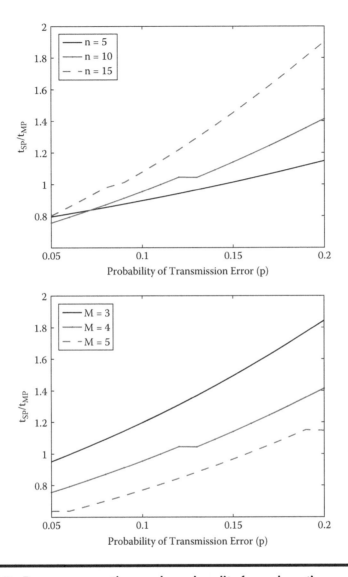

Figure 4.7 **Energy consumption vs. channel quality for each routing mechanism for *M* = 4, *n* = {5,10,15} (left) and *n* = 10, *M* = {3,4,5} (right).**

Next, we investigate the impact of this spatial-variance on the reliability and latency of each protocol, which maintaining the same level of energy consumption. For $M = 5$, $n = 10$, and $\bar{p} = \{0.1, 0.11, \ldots, 0.2\}$, we compare t_{MP} and t_{SP} as a function of the channel for two cases: (i) spatially invariant channel with $p_i = p$ and (ii) spatially variant channel with $p_i = \bar{p} + 0.01$ (*i*3), and the results are shown in the LHS of Figure 4.9. The corresponding results comparing the latency are shown in the RHS of Figure 4.9.

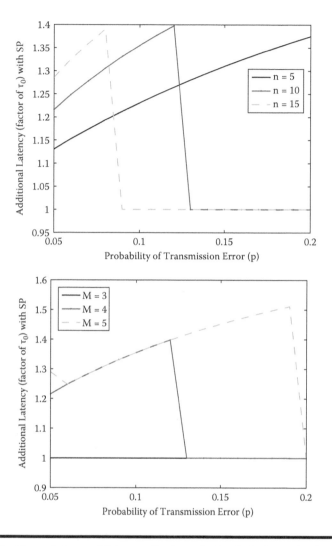

Figure 4.8 Degradation of latency (factor of τ_o) of single-path algorithm vs. channel quality for $M = 4$, $n = \{5,10,15\}$ (left) and $n = 10$, $M = \{3,4,5\}$ (right).

We observe that the spatial-variance in the channel reduces the region of performance gain achieved by the MP protocol in terms of energy efficiency.

4.5 Comparison of Data Delivery Schemes [9]

Relatively few new schemes have been proposed for underwater use, with no single scheme emerging as the obvious choice. In this section, we compare the performance of a simple fixed-path forwarding algorithm in our multipath virtual

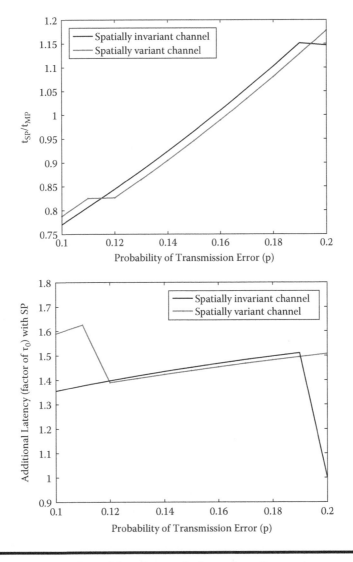

Figure 4.9 Impact of spatial-variance of channel on the energy consumption (left) and degradation of latency (factor of τ_o) of SP algorithm (right) vs. channel quality for *M* = 5, *n* = 10.

sink architecture (which we refer to as MVS forwarding or simply MVS) against the position-based Vector-Based Forwarding [10]. We focus on the data delivery aspect—how data packets travel from sources to the sink—and omit details such as route discovery and query processing.

4.5.1 MVS Forwarding

In MVS forwarding, we employ a simple fixed-path forwarding mechanism [as shown in Figure 4.10(a)] to deliver data from the source to the sinks. All nodes maintain a routing table with the next hop for each sink, using number of hops as the forwarding metric. An initialization period establishes the paths from a node to each of the sinks, in the following manner. Each sink broadcasts a hopcount update message to identify itself. When a sensor node receives this message, it will note the hopcount value (i.e., number of intermediate nodes that are used to forward the message) and rebroadcast the message after incrementing the value by one (hop). Every sensor keeps a record of its hopcount distances from all the sinks, and also the node from which it received the hopcount update (also known as the *previous hop*), during the exchange of hopcount update messages. The previous hop provides information on the path back to each of the connected sinks in the network. Therefore, a node that has sensed data can deliver the packet to any one of the sink nodes that it is connected to, by sending it to the previous hop recursively. This is also known as reverse-path forwarding, a method which is similar to that being used in some terrestrial wireless sensor network routing protocols. These forwarding (reverse) paths are assumed to remain unchanged over time.

4.5.2 Vector-Based Forwarding

Vector-Based Forwarding (VBF) [10], as shown in Figure 4.10(b), can be described as a position-based scheme that employs controlled flooding, with the following features:

Localization. A key assumption made in VBF is that sensor nodes are capable of measuring both the distance travelled by a signal, and its angle of arrival. With this information, nodes can calculate their position relative to the signal source, without global localization information.

Multipath forwarding within a routing pipe. VBF uses a *routing pipe*, a cylindrical region centered around a straight line from source to sink, to deliver data along multiple paths. Only nodes within the routing pipe can forward packets. To achieve this, the pipe's center is attached to packets as a *routing vector*, specifying coordinates of source, sink, and previous hop (all coordinates are relative to the source). The *optimal path* comprises a chain of nodes situated along the routing vector and on the fringe of each others' transmission range.

Variable-path forwarding. Unlike MVS, forwarding paths are not fixed. Rather, every packet transmitted by a node can potentially spawn new paths within the routing pipe.

Adapting to network density. As flooding within the routing pipe may cause contention, VBF attempts to estimate local node density and limit the number of nodes that actually forward packets. Upon receiving a packet, a node

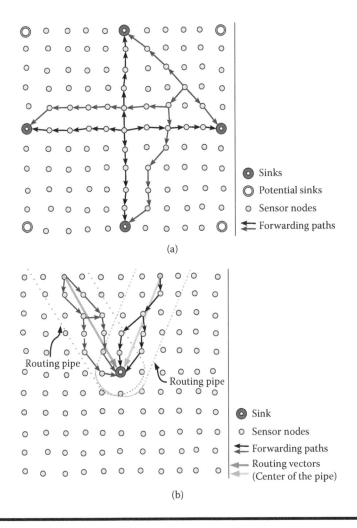

(a)

(b)

Figure 4.10 (a) MVS forwarding. (b) Vector-Based Forwarding. (From Pius W.Q. Lee and Winston K.G. Seah. A Comparison of Two Data Delivery Schemes for Underwater Sensor Networks. *Proceedings of IEEE OCEANS,* June 18-21, 2007, Aberdeen, Scotland. © 2007 IEEE. With permission; and Wiston K.G. Seah and H.X. Tan. Multipath Virtual Sink Architecture for Underwater Sensor Networks. *Proceedings of IEEE OCEANS,* May 16-19, AP, Singapore. © 2006 IEEE. With permission.)

calculates its deviation, or *desirableness factor*, from the optimal path, as a measure of its suitability for forwarding that packet. Nodes nearer to the routing vector and farther away from the previous hop are considered more suitable than those farther from the routing vector and nearer to the previous hop. The packet is buffered for some time before being forwarded; this time interval, or *adaptation delay*, is proportional to the desirableness factor, with zero delay for nodes on the optimal path. During this time, the node may overhear neighboring nodes forward the same packet. It may be useless to forward the packet again, thus requiring the node to compare its desirableness factor relative to its neighbors and decide whether to forward or drop the packet.

4.5.3 Simulations and Results Analysis

We study and compare MVS and VBF through simulations, using the Qualnet simulator [11]. To model the underwater channel, we change several parameters in our simulator: signal propagation speed is set to 1500 m/s, the speed of sound in water; spherical spreading and Rayleigh fading are used to model transmission losses and multipath effects [12]. We model our nodes after the LinkQuest UWM1000 underwater acoustic modem, having a bit rate of 7 kbps and transmission range of 200 m [13]. To avoid drifting away with the currents, underwater sensor nodes must be anchored to the seabed.

In our simulations, we assume that paths to the sinks are set up, that is, the initialization phase for MVS completes successfully, and all sources under VBF are aware of the sink coordinates. To satisfy VBF's key assumptions, we also assume that a signal's angle of arrival and distance travelled are accurately available to nodes. For MVS, sinks are placed along the network boundaries at equal distances from each other; for VBF, a single sink is placed at the center. Some nodes are selected as sources, evenly distributed inside the deployment area, sending 50 packets to the sink. Packets are 128 bytes long. Each source starts transmitting at a random time within 200 s of the simulation start time. A simple CSMA scheme is used at the MAC layer, and all results are averaged over 30 trials.

MVS and VBF have configurable parameters. For MVS, we can change the number of sinks; for VBF, the desirableness modifier a_c and adaptation delay modifier T_{delay}. A lower value for a_c means that nodes must be closer to the optimal path before they will forward packets, and a lower value for T_{delay} reduces the buffer time. In our simulations, we choose parameters as given in the original papers, with a few variants. Altogether, three variants are chosen for VBF (denoted as $VBF_{a_c, T_{delay}}$): $VBF_{0.5,200}$ $VBF_{1.5,200}$, and $VBF_{1.5,600}$, and two variants for MVS: MVS_4 with four sinks, and MVS_8 with eight sinks.

4.5.3.1 Performance Metrics

In our simulations, each packet generated by the sources is assigned a unique sequence number. Since MVS and VBF both employ multipath forwarding, several copies of a given packet may arrive at the sink. We refer to the first copy of a packet arriving at the sink as a *distinct* packet, and subsequent copies as *duplicate* packets.

Furthermore, when evaluating MVS, we regard all sinks as collectively forming a single virtual sink, and a packet is considered as being delivered successfully to the virtual sink when its *first* copy arrives at any one of the physical sinks. We study and evaluate performance of the various schemes using the following set of metrics:

- *Packet Delivery Ratio (PDR)*. PDR is the prime measure of reliability. We define PDR to be the number of distinct packets received successfully at the sink, as a fraction of the total number of packets produced by source nodes.
- *Average end-to-end delay*. The average end-to-end delay is given by the average time taken for a packet to reach the sink.
- *Total number of transmissions*. The total number of transmissions at the physical layer by all nodes provides an estimate of power consumption. This metric should be viewed in relation to PDR, since there is often a trade-off between energy consumption and reliability.

We observe these metrics under varying network density, network size, packet interval, and number of sources.

4.5.3.2 Results: Varying Network Size

We investigate the scalability of both schemes with respect to network size.We vary the number of nodes from 36 to 400, keeping the distance between nodes at 125 m, and scaling the area of deployment proportionately, from 0.625 km × 0.625 km to 2.625 km × 2.625 km. Twenty-five percent of nodes are chosen as sources, sending packets at 180 s intervals.

Figure 4.11(a) shows the PDR for the network. Both MVS schemes show a rapidly decreasing PDR, due to the use of fixed-path forwarding. Packets traversing a fixed path are highly vulnerable to link breaks: the probability of successful delivery decreases exponentially with the number of hops, and a single transmission failure between any two hops totally stops packet delivery along that path. VBF has a slower rate of decrease of PDR because forwarding paths may vary, and a transmitted packet, when received successfully by neighboring nodes, may cause each neighboring node to spawn a new path toward the sink, keeping the packet alive for a longer time.

Figure 4.11(b) shows average end-to-end delay. Delay increases under the VBF schemes because distance to the sink is increased for nodes on the edge of the deployment area, and a fair number of packets from these nodes are still able to reach the sink. For MVS, packets from the center of the network are farthest from the virtual sink and incur the longest delay, but they are also least likely to arrive at the sink; the majority of packets are then received from areas near the sink,

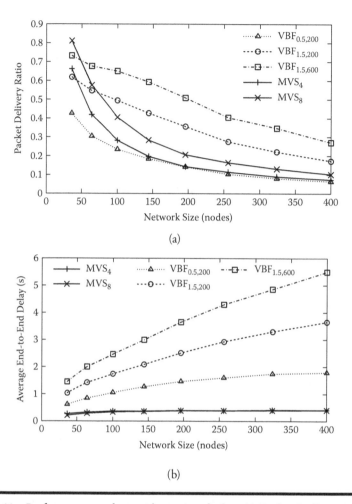

(a)

(b)

Figure 4.11 **Performance under varying network size: (a) Packet Delivery Ratio, (b) average end-to-end delay, (c) total number of transmissions. (From Pius W.Q. Lee and Winston K.G. Seah. A Comparison of Two Data Delivery Schemes for Underwater Sensor Networks.** *Proceedings of IEEE OCEANS,* **June 18-21, 2007, Aberdeen, Scotland. © 2007 IEEE. With permission; and Wiston K.G. Seah and H.X. Tan. Multipath Virtual Sink Architecture for Underwater Sensor Networks.** *Proceedings of IEEE OCEANS,* **May 16-19, AP, Singapore. © 2006 IEEE. With permission.)**

giving MVS a low and constant delay. Both MVS schemes have similar delay because the additional four sinks in MVS_8 are placed at the corners and a packet is more likely to reach sinks along the sides first, rather than sinks at the corners.

The number of physical-layer transmissions is shown in Figure 4.11(c). It increases for all schemes due to more hops being traversed by each packet.

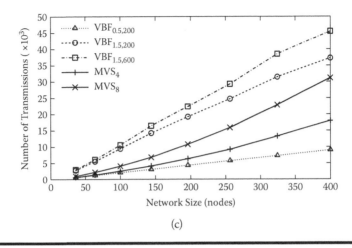

(c)

Figure 4.11 (Continued)

On average, $VBF_{0.5,200}$ expends less power than MVS_4, yet is able to match the PDR of MVS_4 at network sizes beyond 150 nodes. Comparing the schemes with highest reliability, we find that on average, $VBF_{1.5,600}$ requires 2.2 times as much power as MVS_8 to deliver twice as many packets.

4.5.3.3 Results: Varying Network Density

We vary the number of nodes from 36 to 400, confined within a 1 km × 1 km area. Twenty-five percent of nodes are chosen as sources, sending at 180 s intervals. Our results are shown in Figure 4.12.

For all VBF schemes, PDR increases with network density, up to about 150 nodes/km², because more nodes aid in the delivery process. Unfortunately, as shown in Figure 4.12(d), collisions in the network show a great increase, eventually causing a decline in PDR. The problem stems from VBF's density adaptation mechanism: nodes closer to the optimal path can forward more packets and enjoy low adaptation delay; at high node densities, there are too many such nodes that forward packets eagerly, increasing the number of collisions. Due to collisions, neighboring nodes fail to overhear other copies of packets; thinking that no nodes in the vicinity are transmitting, they proceed to forward their own buffered packets, worsening congestion in the area. MVS schemes also have a problem with collisions at higher node densities, but by using diverging paths, collisions increase at a far slower rate and PDR decreases slowly as well.

In [10], delay decreases with increasing node density, but this is achieved by using a high bit rate of 500 kbps, which incurs a drastically lower collision rate. This is not the case in our simulations. Under VBF, congestion along the pipe's center causes successfully delivered packets to be forwarded by nodes near the edge of the routing pipe or closer to the previous hop.

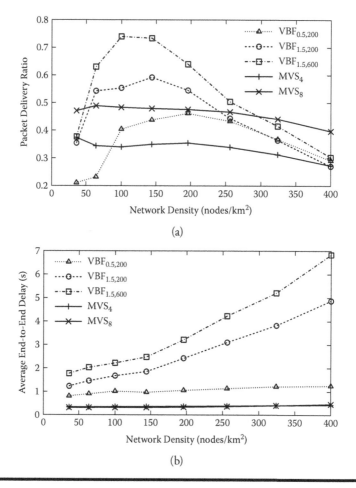

Figure 4.12 Performance under varying network density: (a) Packet Delivery Ratio, (b) average end-to-end delay, (c) total number of transmissions, (d) total number of collisions. (From Pius W.Q. Lee and Winston K.G. Seah. A Comparison of Two Data Delivery Schemes for Underwater Sensor Networks. *Proceedings of IEEE OCEANS,* June 18-21, 2007, Aberdeen, Scotland. © 2007 IEEE. With permission; and Wiston K.G. Seah and H.X. Tan. Multipath Virtual Sink Architecture for Underwater Sensor Networks. *Proceedings of IEEE OCEANS,* May 16-19, AP, Singapore. © 2006 IEEE. With permission.)

Packets are buffered with a longer adaptation delay at these nodes; moreover, increased network density puts more intermediate nodes between source and sink, so packets route through more hops before reaching the sink. As a result, delay increases for $VBF_{0.5,200}$ and $VBF_{1.5,600}$. MVS does not buffer packets and attempts to take the shortest path to each sink; therefore, both MVS schemes incur much lower delay, fairly independent of network density.

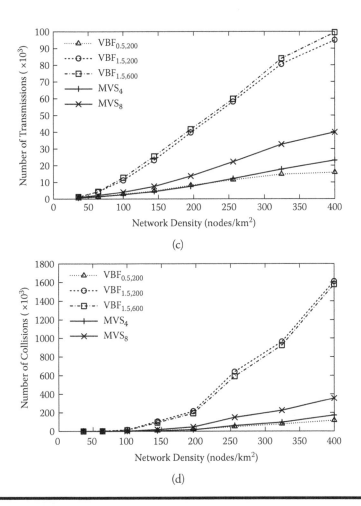

(c)

(d)

Figure 4.12 **(Continued)**

On average, $VBF_{1.5,600}$ consumes about 2.6 times as much power than MVS and delivers 1.2 times as many packets. $VBF_{0.5,200}$ is remarkably different from the other VBF schemes: compared to MVS_4, it consumes similar amounts of energy, but attains higher PDR, comparable delay, and causes a low number of collisions as well.

4.5.3.4 Results: Varying Packet Interval

One hundred nodes are deployed in a 1 km × 1 km area, with 25 sources sending at intervals between 5 s to 180 s. Results are shown in Figure 4.13, using a reversed *x*-axis.

Sending data at smaller intervals introduces congestion into the network. VBF's forwarding paths converge at the sink, making it more vulnerable to congestion

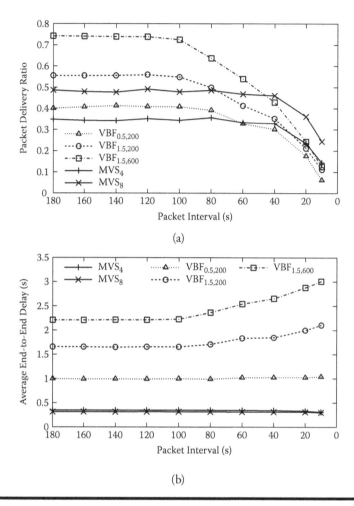

Figure 4.13 Performance under varying packet interval: (a) Packet Delivery Ratio, (b) average end-to-end delay, (c) total number of transmissions. (From Pius W.Q. Lee and Winston K.G. Seah. A Comparison of Two Data Delivery Schemes for Underwater Sensor Networks. *Proceedings of IEEE OCEANS*, June 18-21, 2007, Aberdeen, Scotland. © 2007 IEEE. With permission; and Wiston K.G. Seah and H.X. Tan. Multipath Virtual Sink Architecture for Underwater Sensor Networks. *Proceedings of IEEE OCEANS*, May 16-19, AP, Singapore. © 2006 IEEE. With permission.)

than MVS, which uses diverse paths; consequently, VBF starts to see a decline in PDR from 100 s, while MVS can cope better at low packet intervals, enjoying a stable PDR up to 40 s. Under congestion, packets in VBF tend to traverse suboptimal paths, causing average end-to-end delay to increase. MVS continues to deliver packets with low delay.

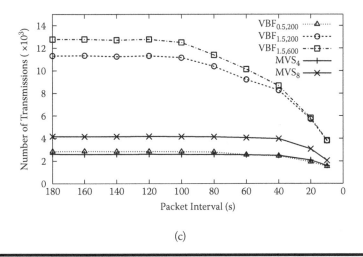

(c)

Figure 4.13   (Continued)

When PDR is stable, $VBF_{1.5,600}$ consumes three times as much power as MVS_8 to deliver 1.5 as many packets. On average, $VBF_{1.5,600}$ consumes 2.6 times as much power as MVS_8 to deliver 1.2 times as many packets. $VBF_{0.5,200}$ expends the same amount of power as MVS_4, yet provides higher reliability for intervals greater than 60 s.

4.5.3.5 Results: Varying Number of Sources

Here again, 100 nodes are deployed in a 1 km × 1 km area, with between 10 to 100 sources sending at 180 s intervals. As shown in Figure 4.14, PDR decreases under VBF as the number of sources increases. It appears that as more nodes transmit, their routing pipes begin to overlap and interfere with each other. The adaptation algorithm was designed for a single routing pipe and is unable to reduce the number of transmissions in the face of increased congestion. Moreover, converging packets cause high contention near the sink. In contrast, MVS maintains a stable PDR. By routing packets through diverse paths, some packets are able to avoid congested areas. Average end-to-end delay increases slightly in VBF since more sources are farther away from the sink. MVS maintains a low and stable delay. On average, $VBF_{1.5,600}$ consumes 2.6 times as much power as MVS_8 to deliver 1.2 times as many packets.

4.5.4 Overall Analysis

In general, we see that VBF attains higher reliability than MVS, at the cost of higher energy consumption. Under a fixed network size of 1 km × 1 km, $VBF_{1.5,600}$ consistently outperforms MVS_8 in PDR by a factor of 1.2 on average and requires 2.6 times as many transmissions. MVS is more scalable with respect to network

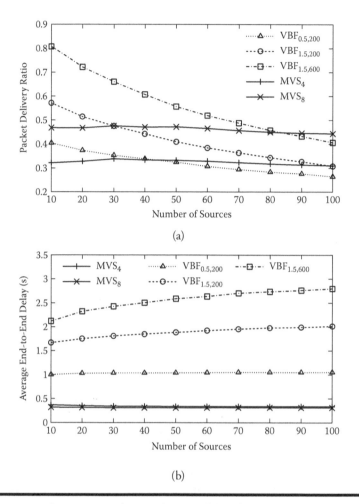

Figure 4.14 **Performance under varying number of sources: (a) Packet Delivery Ratio, (b) average end-to-end delay, (c) total number of transmissions. (From Pius W.Q. Lee and Winston K.G. Seah. A Comparison of Two Data Delivery Schemes for Underwater Sensor Networks.** *Proceedings of IEEE OCEANS,* **June 18-21, 2007, Aberdeen, Scotland. © 2007 IEEE. With permission; and Wiston K.G. Seah and H.X. Tan. Multipath Virtual Sink Architecture for Underwater Sensor Networks.** *Proceedings of IEEE OCEANS,* **May 16-19, AP, Singapore. © 2006 IEEE. With permission.)**

density, packet interval, and number of traffic sources; VBF is more scalable with respect to network size. MVS consistently incurs low delay, while VBF experiences increasing delay in most scenarios.

Given the adverse underwater conditions, some redundant forwarding mechanisms can help to improve reliability, but this aggravates the problem of congestion. MVS and VBF both choose to forward packets through multiple paths. VBF

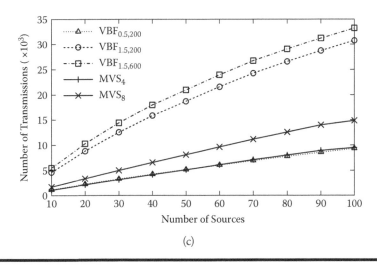

(c)

Figure 4.14 (Continued)

chooses to forward packets through paths near to each other, resulting in more congestion along forwarding paths and high contention at the sink. The approach adopted by MVS—forwarding through spatially diverse paths—proves more effective in avoiding congestion.

Fixed-path forwarding, employed in MVS, is rigid and vulnerable to the intermittent link connectivity under water. Along each of the paths to the virtual sink, a single link break between two hops completely stops packet traversal along that path. VBF achieves higher reliability because the forwarding path can vary; nodes clustered along the forwarding path aid the delivery process, supporting packets on their way to the sink, and attaining higher reliability. An adaptive forwarding scheme for MVS has been shown to provide improved data delivery reliability [14].

Re-transmissions provide a remedy for high transmission losses under water. Considering that sequential re-transmission incurs unacceptable overheads, simultaneous re-transmission is used in MVS to improve reliability and minimize delay, at the cost of higher energy consumption. This is shown in the simulations, with MVS_8 consistently having a higher PDR and number of transmissions than MVS_4.

Buffering packets can improve reliability with a trade-off in higher delay. $VBF_{1.5,600}$ attains the highest PDR across all VBF schemes because it buffers packets for a longer time, allowing the underwater channel to recover before transmitting.

4.6 Applications of Multipath Virtual Sink Architecture

Acquiring accurate telemetry and sensor data on site conditions for underwater construction activities is a big challenge in the offshore engineering community. This technical challenge increases with the drive towards oil production in deeper water with unknown or unstructured environments where the mudline can be

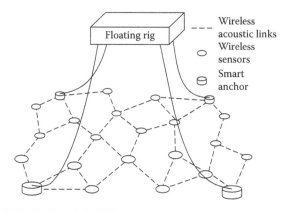

Figure 4.15 Application of MVS architecture for acoustic UWSN in deepwater oil drilling.

characterized as having constantly changing configuration. The multipath virtual sink architecture presented in this chapter is suitable for an acoustic UWSN deployed in offshore deepwater oil drilling, as shown in Figure 4.15. Sensors are deployed on the seabed, and data acquired by sensors is sent to smart anchors by relaying across sensors using multihop communications. These anchors serve as local sinks for the sensor, data, and they are connected by cables to data acquisition system surface platforms where the data is processed and/or forwarded to remote systems for analysis and processing.

More specifically, in deepwater installations, subsea templates (as shown in Figure 4.16), Christmas trees, and manifolds have to be installed accurately in a specified spatial position and compass heading within tight limits, including rotational, vertical, and lateral measurements. The tolerances for a typical subsea installation are within 25 cm of design location and within 2.5 degrees of design heading for large templates and are more stringent for the installation of manifolds into the templates. The resolution of the UWSN needs to be precise to enable accurate installation of the structures in the proximity of other hardware. The signals from the UWSN can be fed back to "intelligent crane hooks" or remotely operated underwater vehicle (ROV) operators who can actively control the positioning of the payload, or be relayed to the bridge of the installation vessel for overall maneuvering. Due to the high daily costs of the crane barge and the marine spread for the installation operation, it is essential that the acoustic positioning system perform with 100% reliability during the brief weather windows in which installations could be safely performed.

Metrological measurements have to be performed after the placement of the structures on the seabed to measure the dimensions between adjacent templates or subsea structures. This dimensional control serves to facilitate accurate fabrication

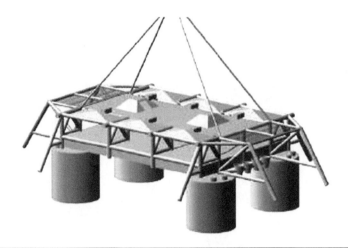

Figure 4.16 Subsea template. (From Vijay Chandrasekhar, Winston K.G. Seah, et al. Localization in Underwater Sensor Networks: Survey and Challenges. *Proceedings of the 1st ACM International Workshop on Underwater Networks.* © 2006, Association for Computing Machinery, Inc. Reprinted by permission.)**

of the connecting spool pieces between the structures. A UWSN that can provide precision measurements for installation can potentially be reused for such metrological purposes. The measurements from the UWSNs can expedite the construction phase by reducing the need for an additional metrology process, resulting in substantial time and cost savings.

The reliability monitoring of mooring systems is another critical requirement in offshore engineering. There are several types of mooring systems employed by the offshore industry for Floating Production Storage and Offloading (FPSO) vessels and other drilling and production vessels. The knowledge of anchor positions can be valuable in the prediction of dynamic behaviors of the mooring systems such as Turret, Single Point Mooring, and Spread Mooring Systems. Vertically loaded anchors (VLA) are currently used in deep water with applications to drilling and operations among other anchors (Figure 4.17). A typical VLA has to be installed at long scope with an uplift angle limit of 15 degrees at the seabed. Also, at the point of maximum holding capacity, further loading results in the anchor being pulled out to the seabed level with decreasing holding capacity. Information pertaining to the position and inclination of the anchors relative to the vessel will enable consistent monitoring of system reliability, safety, and optimization.

Besides the offshore engineering industry, similar multi-sink architectures can be applied to other applications such as tsunami early warning systems, environmental monitoring of the oceans, and perimeter security of naval and other key installations.

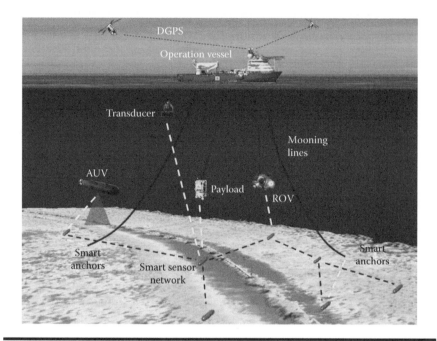

Figure 4.17 **Reliability monitoring in deepwater mooring system. (From Vijay Chandrasekhar, Winston K.G. Seah, et al. Localization in Underwater Sensor Networks: Survey and Challenges.** *Proceedings of the 1st ACM International Workshop on Underwater Networks.* **© 2006, Association for Computing Machinery, Inc. Reprinted by permission.)**

4.7 Conclusions and Future Work

Underwater acoustic networks are envisaged to be the enabling technology for oceanographic data collection, pollution monitoring, offshore exploration, and tactical surveillance applications. Although reliable and efficient communication protocols are in place for terrestrial networks, they cannot be directly employed in the underwater environment due to the unique characteristics of underwater acoustic channels such as higher bit error rates and larger propagation delays.

As such, we have proposed the use of a virtual sink architecture, in which sensor nodes can forward data to one or more spatially diverse sinks to avoid contention and achieve high reliability despite the adverse network conditions. We have used analysis and extensive simulations to compare the performance of our approach against that of conventional single-path single-sink with a finite number of permissible re-transmissions, as well as known data delivery schemes for underwater networking. From our results, we have shown that it is indeed possible to achieve better network performance using our proposed virtual sink architecture. In parallel, we have also developed data delivery schemes specifically designed for this architecture, using packet-cloning techniques [15] and adaptive re-routing [14] to address the dynamic underwater channel conditions.

The high error rates lead to significant packet losses and this makes the design of efficient ARQ schemes to ensure reliability very important. In this aspect, an opportunistic acknowledgment scheme suitable for Stop-and-Wait ARQ protocols has been proposed [16] and shown to offer better latency and energy efficiency than traditional non-opportunistic schemes for both one- and two-dimensional multi-hop acoustic channels. The scarcity of bandwidth also compels us to pay special attention to interference management in medium access techniques for underwater communications [17].

Looking ahead, it remains a challenge to coordinate access (i) between navigation and data signals and (ii) between Autonomous Underwater Vehicles (AUV) and sensor nodes due to the limited underwater bandwidth. In this aspect, we have started exploring how dynamic spectrum sharing concepts inspired by the advance in cognitive radio technology can be used for spectrum management to achieve integrated communication and navigation in integrated underwater acoustic systems [18].

A necessary aspect of underwater sensor networks research is the validation of ideas in a real deployment scenario, and it is our ongoing effort to implement the proposed schemes in prototypes and test them in actual underwater environments.

Last, but not least, there is a critical need to address realistic scenarios for sensor network deployments. Most of the scenarios that have been considered so far tend to be limited and even unrealistic. For example, how often are sensors evenly or uniformly distributed over the area of deployment, especially when they are dropped in large number from the sky?—a quintessential assumption in many proposed wireless sensor network protocols and algorithms.

References

1   M. Souzer, M. Stojanovic and J.G. Proakis, Underwater Acoustic Networks, *IEEE Journal of Oceanic Engineering*, 25 (1), 72–83, (Jan 2000).

2. A. Das and D. Dutta, Data Acquisition in Multiple-Sink Sensor Networks, *Mobile Computing and Communications Review*, 9 (3), (July 2005).

3. J. Catipovic, Performance Limitations in Underwater Acoustic Telemetry, *IEEE Journal of Oceanic Engineering*, 15 (1), 205–216, (Jul 1990).

4. D.B. Kilfoyle and A.B. Baggeroer, The State of the Art in Underwater Acoustic Telemetry, *IEEE Journal of Oceanic Engineering*, 25 (1), 4–27, (Jan 2000).

5. M. Haenggi and D. Puccinelli, Routing in Ad Hoc Networks: A Case for Long Hops, *IEEE Magazine*, 93 (10), 93–101, (Oct 2005).

6. Winston K.G. Seah and H.X. Tan, Multipath Virtual Sink Architecture for Underwater Sensor Networks, *Proceedings of the MTS/IEEE OCEANS 2006 Asia Pacific Conference*, May 16–19, 2006, Singapore.

7. Winston K.G. Seah and H.P. Tan, Multipath Virtual Sink Architecture for Wireless Sensor Networks in Harsh Environments (invited paper), *Proceedings of the First International Conference on Integrated Internet, Ad Hoc and Sensor Networks (InterSense 2006)*, May 29–31, 2006, Nice, France.

8. S. Mueller, R.P. Tsang and D. Ghosal, Multipath Routing in Mobile Ad Hoc Networks: Issues and Challenges, *M.C. Calzarossa and E. Gelenbe (Eds.): MASCOTS 2003, LNCS 2965* (2004), 209–234, (Springer-Verlag Berlin Heidelberg).

9. Pius W.Q. Lee and Winston K.G. Seah, Comparison of Data Delivery Schemes for Underwater Sensor Networks, *Proceedings of the OCEANS 2007 Europe Conference*, 18–21 June, 2007, Aberdeen, Scotland.

10. P. Xie, J. H. Cui and L. Lao, VBF: Vector-Based Forwarding Protocol for Underwater Sensor Networks, *F. Boavida, T. Plagemann, B. Stiller, C. Westphal and E. Monteiro (Eds.): NETWORKING 2006, LNCS 3976* (2006), 1216–1221, (Springer-Verlag Berlin Heidelberg).

11. Scalable Networks Inc., Qualnet, http://www.scalable-networks .com

12. P.C. Etter, *Underwater Acoustic Modeling and Simulation*, 3rd edition, Spon Press, New York, 2003.

13. LinkQuest Inc., UWM1000 Underwater Acoustic Modem, http ://www. link-quest. com/html/uwm1000.htm

14. P. Sun and Winston K.G. Seah, Adaptive Data Delivery for Underwater Sensor Networks, *Proceedings of the MTS/IEEE OCEANS 2007 Americas Conference*, Sep 29–Oct 4, 2007, Vancouver, Canada.

15. P. Sun, Winston K.G. Seah and Pius W.Q. Lee, Efficient Data Delivery with Packet Cloning for Underwater Sensor Networks, *Proceedings of the International Symposium on Underwater Technology*, Apr 18–20, 2007, Tokyo, Japan.

16. H.P. Tan, Winston K.G. Seah and L. Doyle, A Multi-Hop ARQ Protocol for an Underwater Acoustic Channel, *Proceedings of the OCEANS 2007 Europe Conference*, 18–21 June, 2007, Aberdeen, Scotland.

17. H.P. Tan, C. O'Sullivan and Winston K.G. Seah, Interference Management for Medium Access Control in CDMA Underwater Acoustic Sensor Networks, *Proceedings of the IEEE 67th Vehicular Technology Conference (VTC2008-Spring)*, May 11–14, 2008, Marina Bay, Singapore.

18. H.P. Tan, Winston K.G. Seah and L. Doyle, Exploring Cognitive Techniques for Bandwidth Sharing in Mobile Underwater Networks, *Proceedings of the OCEANS'08 MTS/IEEE KOBE-TECHNO-OCEAN '08 (OTO'08) Conference*, Apr 8–11, 2008, Kobe, Japan.

19. Vijay Chandrasekhar, Winston K.G. Seah, et al., Localization in Underwater Sensor Networks: Survey and Challenges, *Proceedings of the 1st ACM International Workshop on Underwater Networks*, Sept. 25, 2006, Los Angeles, CA.

FAULT TOLERANCE AND TIME SYNCHRONIZATION

III

Chapter 5

A Survey of Fault Tolerance in Ad-Hoc Networks and Sensor Networks

Songqing Yue, Yang Xiao, Xihui Zhang, Jiming Chen, Jianhui Zhang, and Youxian Sun

Contents

An ad-hoc network or sensor network contains stationary or mobile nodes communicating with each other via wireless media. Components in a network will eventually fail, which leads to unanticipated disruptive failure behaviors or even service unavailability. Therefore, systems need to be designed as fault tolerant so that they will function continuously to conduct their tasks even though component failure occurs. Fault tolerance has been well studied in operating systems and distributed systems. Most of the traditional methods resort to adding redundancy.

In this chapter, we first introduce some basic concepts and design issues that are related to fault tolerance in ad-hoc networks and sensor networks. Then we provide a survey of the problems and faults that can occur in these networks and the corresponding fault-tolerance techniques to counteract the problems and faults. Finally, we review and discuss in detail the consensus problems in fault-tolerant distributed systems.

5.1 Fault Tolerance in Networks

Making networks fault tolerant is important due to the following reasons: (1) a single switch failure in a wired fiber network may cause a huge loss of traffic because of the large capacity [1]; (2) due to the mobility of devices that offer services in an ad-hoc network, a user may not be able to depend on a special device [2]; (3) network services are vulnerable to malicious attacks; (4) the number of errors in a software grows with the augmenting of complexity and scale of the software [3]; (5) in sensor networks, sensors may be subject to failures because of some environmental parameters like temperature, pressure, motion, etc. [4].

Survivability or fault tolerance refers to the ability of a system to continue fulfilling its predefined task in time even though some components in the system fail due to accidents or attacks [5]. Survivability deals with trade-offs among software quality attributes determined by the mission [5]. Survivability solutions provide risk management strategies, and a standby strategy is needed in case of failure of an existing system [5].

A graph can present a network with vertices and edges indicating respectively nodes and links in a network. Edge connectivity is defined as the number of the fewest edges or nodes to remove in order to make the graph disconnected [1].

Hence, if edge connectivity equals k, after removing $(k-1)$ edges, the network will remain connected. Edge disjoint paths are those that have no edges in common. Connectivity is used to measure survivability in common [1].

5.2 Design of a Survivable Ad-Hoc Network

In [1] a disaster network design is studied, and the aim of the design is to create relay points based on positions of a header, targets, and the radio radius. The design of survivable ad-hoc networks has to satisfy the survivability requirements. However, due to different properties in the location and dependence and broadcast features, it is difficult to build a model for wireless ad-hoc networks with conventional wired network techniques [1]. Unlike wired networks, where networks have topological freedom, the propagation strength of wireless networks strongly depends on mutual locations of the nodes [1]. In a wired network the transmission is single dimensional just along the cable, while in a wireless network it transmits in three dimensions of the space [1].

In the network model presented in [1], nodes and spheres are represented to stand for nodes and links. The model is simplified as a two-dimensional plane, where radio supply area is considered to be a circle rather than a sphere, and circles are further modified to polygons [1].

By following adjacent polygons on a polygon grid, paths can be found from the source to targets [1]. A dual graph is used in planar graph problems to find paths from the source to targets by exploiting dual vertices and edges [1]. A dual edge connects dual vertices to adjacent areas and intersects area boundary only once [1]. Dual vertices of adjacent areas are connected by dual edges, and dual edges follow adjacent vertices of a polygon grid on the mesh of a dual graph. A heuristic algorithm is used to find appropriate paths [1].

5.3 An Architecture for Survivable Wireless LANs

Resiliency and survivability are important features for wireless LANs (WLANs) since access point (AP) central authority with centralized control is especially vulnerable to attacks [7]. Upon AP failure, the nodes under this AP lose network connectivity.

The scheme, called IAMS, provides a certain degree of survivability against failures and attacks [7]. In IAMS, nodes of the failed AP switch from an infrastructure mode to an ad-hoc mode to regain network connectivity after the AP failure; if a node detects that the AP it connects to fails, the node will change to an ad-hoc mode from an infrastructure mode to get network connection [7].

Nodes can be categorized into several types, such as a bridge node, a leader node, and an ordinary node [7]. A bridge node refers to the node in the radio range of several APs but only connected to an AP. Those bridge nodes can help to regain connectivity of a network by serving as a bridge to relay the traffic of nodes linked to a failed AP to other APs [7]. The role of a leader node is to function as a control head distributed in the network [7]. The duty of a leader node involves deciding routing paths, isolating corrupt nodes, as well as undertaking the duty of controlling other nodes after an AP fails [7]. Therefore, the leader nodes play an important part in survivability of a network. They handle malicious nodes that have caused AP failures. Some known detection schemes can be utilized to identify malicious nodes that have resulted in failures of APs or are planning to compromise services.

Malicious nodes can be removed transparently by removing the links of the nodes towards other nodes in the network in a distributed manner. In this case, no information needs to be exchanged among nodes, which is very important to maintain the integrity of networks. It is also indispensable in the following cases:

- Instead of crashing a network completely, a compromised node may choose to take advantage of the weak points in the network to eavesdrop [7].
- After a malicious node finds out the weak points of a network, it may mount various attacks [7].

Once the malicious or compromised nodes are identified, the leader nodes will remove them independently and transparently without first communicating with other nodes [7]. Removing nodes may result in disconnection, so that it is necessary to recalculate routes for each node to bridge nodes.

The IAMS scheme is based on the following assumptions:

- First, they assume that at a given time only one node is compromised and each node possesses a key pair that enables secure communication between nodes in one hop distance [7].
- Second, they assume that in the network, at least one node is in the radio range of multiple APs [7].
- The last assumption made is that APs are devices equipped with some detection or monitoring software, which is intended to make any detection- or monitoring-related decisions [7].

Three major components of IAMS are standards of leader selection, graphs of topology, and mode switching technique [7]:

- *Standards of leader selection*: The following are the two criteria used to select a leader:
  - *Trust-based selection*: If a node is regarded to be more trustworthy than other nodes, this node can be selected as a leader [7]. Trust-based criterion has been widely used in network areas [7]. The advantage of this scheme lies in that the trusted leader chosen is less likely to become malicious than other nodes.

- *Region-based selection*: A leader node is selected according to its closeness to others and its connectivity towards those bridge nodes [7]. The ideal case may be that all leader nodes are equally distributed in the network and also they are close to bridge nodes connected to them, so that the network can perform optimally with traffic distributed equally [7].

■ *Topology graphs (TG)*: Topology graphs are the routes used by leader nodes to route data when an AP fails [7]. TGs are represented with two parameters: network graph and adjacency list, and are built up by APs when the network is in a normal condition [7]. The TGs are sent periodically to those leader nodes only. This is helpful to reduce overhead and is necessary to keep compromised nodes from learning the routing information [7].

■ *Mode switching technique*: The mode switching technique includes three phases:
  - If nodes cannot receive beacon messages from the AP for a defined time period T, the nodes will assume that the AP has failed, and then they will switch themselves to the ad-hoc mode [7].
  - In the ad-hoc mode, the leader nodes will take the responsibility of maintaining the continuous function of those nodes by sending control packets to other nodes [7]. The control packets contain routing details like routes, traffic load, as well as updating information caused by the removal of corrupt nodes [7].

■ After receiving routing instruction from leader nodes, each node begins to route packets in the ad-hoc mode.

5.4 Dynamic Task-Based Anycasting

5.4.1 Introduction

The paper in [2] focuses mainly on higher layer issues concerning the application modeling in a mobile ad-hoc network (MANET) environment. The authors argue for the design of a distributed application framework [2]. The framework is based on task graphs to enable various applications based on the feature of resource discovery in a MANET [2]. In the framework, a distributed application can be described as a whole task that is composed of several sub-tasks [2]. Those sub-tasks are supposed to be executed on multiple nodes, which can be classified with particular roles [2]. When performing a specific task in a MANET, data flows between different types of devices, which causes dependencies among those devices [2]. To logically represent these dependencies, a task graph of the application is exploited in the paper [2].

To tackle the problem of performing tasks distributed over a MANET, the paper proposes an efficient algorithm of "dynamic discovery and selection of particular devices" [2], which are sufficient to execute the tasks, from multiple

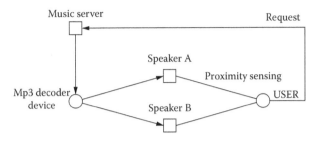

Figure 5.1   Stereo music service. (After P. Basu, W. Ke, and T.D.C. Little, *Mobile Networks and Applications,* **8(5), 593-612, 2003.)**

candidates with the same functionality. In effect, this is achieved via the task graph representation of applications and this process can be referred to as Dynamic Task-Based Anycasting [2]. The algorithm is capable of detecting failures in a MANET by monitoring the logical links between devices and recovering from those failures by substituting the malfunctioned devices for other devices with similar capabilities [2].

A task graph is proposed in the paper [2] to represent the abstraction of higher-level applications, which can be divided into several sub-tasks. A task graph consists of two elements: edges and nodes, in which nodes stand for the classes of devices and services that are necessary to execute a task, while edges indicate the relation between those nodes to performing the task [2]. Figure 5.1 is an example provided in the paper [2] showing the TG in a music service application, where a speaker can be selected automatically based on the physical location of the user.

For each type of device in TG, one suitable instance needs to be chosen to take part in task execution and this process is called Dynamic Task-Based Anycasting [2]. To execute a task, we need to choose one proper device in each type of devices. The process of making the selection of a proper device is called Dynamic Task-Based Anycasting [2], in which the word "dynamic" indicates that the device selected to function might change because of mobility [2].

The merits of the proposed algorithm are explained as follows [2]: If the device that is performing a task fails, a new device with similar function will be chosen to go on with the task. Thus the algorithm can adapt to changing circumstances and dynamically rebuild failed segments of a TG whenever disruption happens [2]. This implies that the model works if only one device in each type of device with the ability to execute a special sub-task is available in a MANET [2].

5.4.2 The Framework Based on TG

The original adoption of a task graph was primarily in the field of parallel and distributed computing, where a complicated task can be divided into multiple sub-tasks and be assigned to several homogeneous devices connected with high-performance links to reduce the whole execution time [2]. However, the task graph concept in the paper [2] concerns only several heterogeneous devices, which can provide specialized services and connect with each other. To some extent, the task graph scheme here enjoys more generality than the one used originally [2]. Also, the algorithm differs from previous research in that it works at the logical layer [2]. The logical layer is on the top of the service discovery, which is widely researched in the area [2]. Graph theoretic approaches are considered as a promising and effective tool to model distributed applications in both traditional distributed systems and in mobile distributed platforms [2]. Some terms to be used in the paper [2] for clarifying the scheme are defined as follows.

According to the paper [2], a *device* in this work means a physical entity with the capacity of computing and communicating with other devices and its surroundings. An ordinary device is usually with several functioning components such as an embedded processing unit, sensors, a wireless communication module, as well as user interfaces. A device can be a *specialized device* if it offers only a particular service, or a *multi-purpose device* with the ability to serve more than one purpose. The capability of one device can be found in its *attributes*, which includes both the static and the dynamic properties, for instance, the resolution of a digital camera, the location, power level of a sensor, etc.

The word *service* is used to describe the functionality that a device can provide. A *node* represents a device or a group of devices that are equipped with the smallest set of features to provide a specialized service. The primary feature of a device can be referred to as *class* or *category* or *type*. An *edge* indicates the connection between nodes with attributes that are sufficient to complete a task, for example, weight, data rate, allowable bit error rate, physical proximity, etc.

In the following sections, we first show the theoretical foundation of the scheme and then discuss some issues about the task embedding. Then we focus on the introduction of a realizable distributed algorithm that can efficiently implement a tree task graph in dynamic, mobile environments.

5.4.3 Task and Task Graphs

A task refers to the work that is performed by a device to output certain results [2]. As part of a big task, a sub-task might be performed by parts of all devices. If a task cannot be divided any longer, we call it an atomic task, which is performed by a single device [2].

A task graph can be represented by a graph donated by $TG = (V_T, E_T)$. Here V_T indicates the group of nodes that are required to cooperatively perform a task, and E_T stands for the group of edges indicating data flow between involved nodes [2]. The following example shown in the paper [2] aims to help make those terms clear and understandable As shown in Figure 5.2, a laptop that is able to, among other services that it can offer, convert a PDF file into a printable format and a printer that is used to print some pages are connected with each other to form a complex node to offer the service of printing.

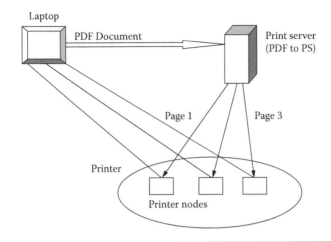

Figure 5.2 A smart printing service. (After P. Basu, W. Ke, and T.D.C. Little, *Mobile Networks and Applications,* **8(5), 593-612, 2003.)**

5.4.4 Tuple Architecture for Data Flow

The primary attribute of a node or a device can be referred to as a class. Basically, a group of tuples is used to represent the requirements of tasks with respect to data flow between devices belonging to different classes [2]. Every physical device performing a task has such a group of tuples, each of which is associated with a logical unit of data flow between different distributed elements of an application [2].

Application data flowing between nodes can be shown by a tuple architecture. Assume that a device of class A gets data from nodes in class X, Y, then sends data to other nodes in classes Z. Then this data flow can be represented as the tuple A: $[X, Y; Z]$. The reason for using a tuple architecture to represent the task processing is because (1) using a standardized language, the data flow of applications can be predefined and saved in a file, from which a TG can be easily generated, and (2) the tuple architecture can control the flow of actual data after embedding the task graph on the MANET [2].

5.4.5 Instantiation of TG Nodes by Using a Distributed Algorithm

Before performing a task, a group of appropriate devices that possess the required attributes needs to be discovered in the network. From these devices, a set of suitable ones needs to be selected to execute the particular task [2].

The main goal of the algorithm defined in [2] is to embed a task graph onto a MANET with the purpose of enhancing performance. The assumption followed by the algorithm for the heterogeneous devices is that each device can only offer a particular service and each node is a simple one [2]. Another assumption is we can take advantage of existing routing and transport protocols like DSR or TCP [2].

The embedding process begins when the application layer of a node submits a TG to its TG layer [2]. Before that it is not needed to exchange the state of devices in the network. The instantiation process is totally on-demand. The copies of the same algorithm are executed by all devices in the network [2]. All devices in the system exist in a state s such that $s = S1 \times S2$, where $S1$ = {COORDINATOR, NON_COORDINATOR} and $S2$ = {UNINSTANTIATED, WAIT_FOR_ACK, INSTANTIATED, SUBTREE_INSTANTIATED} [45]. We refer to a device as a "coordinator" when it plays the part in the job of instantiating a small group of nodes in the TG [2].

Usually the embedding process starts from a User node U with a distributed search through the network G simultaneously with a Bread-First Search (BFS) through a task graph [2]. The spanning tree deduced on TG with the root of U is referred to as a BFS-tree of TG (BFST$_{TG}$) [2].

A greedy solution is proposed to keep the performance of instantiation (the term dilation is used to express it in this paper) [2]:

■ The first step of the algorithm begins from U when it tries to progressively map the nodes of BFST$_{TG}$ to devices that are comparatively near and the edges to paths that are relatively short in G.

■ The search proceeds along the nodes of BFST$_{TG}$ in a distributed way until the instantiation of any two nodes A, B cannot influence one another if A is not the parent of B in BFST$_{TG}$ [2].

■ The algorithm instantiates nodes with greed in TG and explores only the space around a device that has been instantiated to instantiate another node [2].

The crucial phases of the scheme in [2] can be described as follows: The user node U sends search queries simultaneously to each TG neighbor node (here X and Y). The search packet contains the query fields, which include the principal attributes of nodes X and Y, respectively. A packet is rebroadcasted by a node Z in the TG layer (1) if Z cannot match the type of searching query, (2) if the "time-to-live" (TTL) does not equal zero, and (3) if Z has not received the same query before [2].

After getting a search packet, a free instance Y_j of type Y sends a packet containing a candidate query to U to show that it is ready to join the task [2]. Then the state of Y_j will be changed to WAIT_FOR_ACK and after checking Y's state, and U sends an ACK to Y_j if Y still has not been instantiated. The duty of a coordinating device is to accept or reject responses from candidate devices before the instantiation is finished. The first candidate device responding to a searching query is chosen as the instance of type Y in the paper [2]. This is a simple approach followed by selecting nearby devices. More sophisticated criteria of the selection process, like maximizing the energy efficiency and remaining battery lifetime, can also be adopted [2].

In the ACK U has sent to Y_j, a task graph is included that shows all device types in a sub-tree with the root at Y. Other than sending an NACK to all other instances of Y from U, a timer is set up by Y_k immediately after it sends a candidate query to U. The state of Y_k will be changed back to UNINSTANTIATED if the timer set has expired before Y_k gets the ACK. After getting an ACK from U, the state of Y_j will be changed into (INSTANTIATED, COORDINATOR) and Y_j will send a CN_CONFIRM to U to confirm its role to U. By now, Y_j undertakes the responsibility of a coordinating node for those with its parent Y in the TG needing to be instantiated. In the next step, Y_j becomes a local coordinator whose responsibility is to instantiate all its child nodes in the sub-tree and the same instantiating process goes further [2]. Any device functioning as a leaf node in $BFST_{TG}$ sends sub-tree confirmation along with the conformation packet because it does not need to send a further search request [2].

When sub-tree confirmations are received from all instances of child nodes, the Y_j also sends a sub-tree confirmation to its BFS parent with which an instantiated sub-TG is included [2]. By now the entire instantiation process of the sub-tree at Y_j is complete [2]. After receiving sub-tree confirmations from all its child nodes, U begins to conclude the embedding process by sending out the tuples and instantiated TG to those selected devices so as to let all node instances learn the addresses of other instances in TG [2]. The user application starts data transmission after the instantiation information is exchanged. The tuples take the responsibility of governing the corresponding data flow during the future data transmission.

5.5 Byzantine Fault Tolerance—A Practical Approach

5.5.1 Introduction

The paper in [3] proposes a practical Byzantine Fault Tolerance algorithm working in asynchronous environments like the Internet that is capable of securing the safety and correctness as long as at most $(n − 1)/3$ in totally n nodes are instantaneously faulty. The algorithm employs only one message travelling around all nodes to perform operations that are read-only and two messages for operations that are read-write, which is quite efficient to be immune to a denial-of-service attack [3].

5.5.2 The Algorithm Description

In the algorithm, services are represented as a state machine that is replicated in several nodes in the distributed system [3]. Each node with the state machine keeps the state of services and is capable of executing the services [3]. To each replica, an integer from 0 to N is assigned, and to make things simple, an assumption is made that N is equal to $3m + 1$ where m refers to the number of replicas at most that might be malicious [3]. Some terms used to clarify the algorithm are defined as follows. According to the paper [3], *Views* refer to a series of configuration through which replicas move. A particular replica is selected to be the *primary* that acts like a leader, while the others all serve as *backups* [3]. Views are indicated by consecutive numbers, and view changes may be performed when the primary has been detected to fail [3].

The algorithm can be described with the following steps [3]:

A client sends a message to the *primary* to request the operation of a service.

The *primary* then broadcasts the request to all the backups [3].

After receiving the request, replicas perform the service according to the request and then send the result back to the client [3].

The client cannot make sure of the correctness of the result until it receives $m + 1$ replies and all $m + 1$ ones are same [3].

Two requirements imposed on replicas, as with all state machine replication techniques, are that they must be deterministic and that they must start in the same state [3]. In other words, with a given set of parameters and in a given state, the operation execution must have the same outcome. If these two requirements are met, then the algorithm can secure the safety feature by making sure that all non-faulty replicas may come to an agreement to fulfill the requests even if there are failures [3].

5.6 Stochastic and Designed Faults—Merging Paradigms of Survivability and Security

5.6.1 Introduction

System faults have always been a great concern for both security and fault-tolerance communities. However, the approaches that the two groups of researchers use to model and correct these faults are much different. A system can be viewed as survivable by one community but not by the other. That's why both approaches of two different communities cannot be comprehensive [8]. People from the security community study faults with respect to "statistically dependent events" caused by hard intruders, whereas the fault-tolerance research asserts that those faults are stochastic and can be represented as "random variables with probability distributions" [8]. Under the circumstances, the paper in [8] proposes a new paradigm aiming for understanding the survivability of a given system in a comprehensive manner.

Below are some definitions and concepts from [8] that are helpful to understand the conceptual difference between two different areas.

■ Attack—This is defined as a malicious action fault that aims to break some security properties with intention.

■ Vulnerability—This is defined as a fault that is accidental or intentional that can be taken advantage of to mount an attack [8].

■ Intrusion—This is defined as an attack that is successfully launched with malicious intention [8].

Based on conventional security research, the above three properties can be defined with respect to a common security feature like integrity, availability, or confidentiality [8]. For instance, there may be an availability attack or a confidentiality intrusion [8]. A system that can tolerate confidentiality intrusions may not withstand an availability intrusion [8]. We can illustrate this using the following example: A system with redundant copies of a data item X can tolerate availability intrusions as long as not all copies of X are damaged by the intrusion [8]. Whereas, the same system may not tolerate a confidentiality intrusion because a confidentiality intrusion may cause an unauthorized read of X rendering the confidentiality of X being violated [8].

5.6.2 Two Types of Faults

An intruder, based on its characteristics, can be categorized as a hard intruder or a gremlin [8]. Hard intruders refer to those who have "relatively high-value objectives, low risk aversion, high skills, and high resource levels" [8]. Hard intruders may constitute a team working together to obtain a very high value objective, and the reoccurrence rate of attacks from hard intruders is high. They cannot model those attacks correctly with Byzantine faults for that they influence on components with statistical dependence [8].

In contrast, gremlins have "no specific objective, low skills, low risk aversion" [8]. The most important fact is that gremlins can start attacks to any components at any time [8]. Gremlins are not real beings so it is difficult to stop them using existing techniques. It seems impossible to stop a gremlin using trusted design and development since gremlins are spontaneous intruders with no specific objectives [8]. The impact on different components seems to be statistically independent, so Byzantine faults are good models to describe the action of gremlins [8].

5.6.3 Problematic Faults

The fault-tolerance communities focus mainly on stochastic faults while the security communities on designed faults. These two classes of faults are illustrated in detail as follows.

5.6.3.1 Designed Faults

Designed faults refer to those faults caused by hard intruders and deliberately devised to invalidate some assumptions and assertions about the system under attack [8]. The following examples of faults all fall into the category of designed faults: common node faults that repeatedly attack redundant components in order to destroy the redundancy, and architecture faults that are designed to attack the weak points of a system [8].

Designed faults are the main concern of the security field, while the fault-tolerance communities overlook them, because these faults do not seem to have impact on the tolerance of a system in an independent manner [8]. Redundancy-based approaches work only under the assumption that the attacks are not replicated in the same way and the reconfiguration-based methods work on the condition that the attacks do not restructure to adapt to new status [8]. However, designed faults by their characteristics are capable of exploiting the exact same avenues. One of the limitations of fault-tolerance techniques lies in that the approaches used assume absolutely random behavior, so they only focus on faults that act unpredictably [8]. Therefore, it seems inappropriate to use stochastic variables to model those designed faults by employing fault-tolerance techniques.

5.6.3.2 Stochastic Faults

Stochastic faults are committed by gremlins. Many accidental reasons, like software or hardware flaws or physical damage, may serve as the cause of stochastic faults [8].

Redundant fault containment regions are employed by fault-tolerance approaches to tackle stochastic faults, which are proven to be of great power [8]. However, approaches of security that are used to withstand designed attacks are designed on the basis of hard intruders. Following each class of fault, there is a process from design to deployment. The main purpose of this is to constitute a system with a few trusted components while others are entrusted [8]. The trusted components refer to those that the hard intruders cannot access [8]. Second, the hard intruders cannot succeed in manipulating any combination of entrusted components due to the interaction between those trusted components [8]. Based on the view of a trusted component, the attackers behind random faults are not real and thus they are not worthy of consideration [8]. Therefore, no measure has been taken to deal with those stochastic intruders.

5.6.4 The Paradigm Shift

Any intrusion-tolerant or survivable system based on redundancy but not taking into account the hard intruders may appear weak when facing designed attacks [8]. Similarly, any system that is based on trusted components but ignores stochastic faults may also seem to be powerless against attack from gremlins [8]. In order to survive in both scenarios, a shift method is necessary to set up really survivable systems [8]. Here are the three ways proposed to achieve this goal [8]:

■ Make fault-tolerance approaches effective in dealing with designed faults. The changes proposed lie in that the approaches should indicate due trust between redundant components, and those redundant components are able to gain the required trust level.

■ Make changes to trusted-component approaches to adapt to stochastic faults. The changes proposed should be based on trusted-component approaches and also should offer solutions for tackling stochastic faults by the means of redundancy and reconfiguration.

■ Increase the impression of models like stochastic process algebra to constitute practical systems (the stochastic process algebra is elaborated in the paper [8]).

5.7 Fault-Tolerance Techniques

5.7.1 Cache Management of Dynamic Source Routing for Fault Tolerance in Mobile Ad-Hoc Networks

5.7.1.1 Introduction

Mobile nodes in ad-hoc networks use wireless media to communicate. Two nodes can communicate with one another either when both nodes are in each other's transmitting range or through a group of intermediate nodes that can form a continuous chain of nodes with each node in the transmission range of its neighboring nodes and every node is willing to transmit the packet [9].

There are a number of routing protocols available for routing the packets through ad-hoc networks. Among those protocols, dynamic state routing is one in which routes only exist between nodes that are in need of communication, and route caching is kept in order to reduce route discovery overhead [9]. But managing the cache is one challenging problem as the nodes are not stationary and it is possible that the nodes may not be in each other's transmission range when the route is requested again after some time [9]. To deal with this problem a cache management protocol is proposed in [9] with the goal to avoid replying with an obsolete route from the cache, manage to maintain the correct routes in the cache, and indicate the dynamic network changes using a local scheme. A node uses the signal strength, denoted by the parameter of signal strength threshold, to determine the stability of the neighboring nodes that exchange packets with it [9]. When received packet signal strength falls below the threshold, the packet's sender is considered unstable. The receiving node also marks all the routes through that sender as stale [9]. The case management protocol is explained in detail in the following sections.

Confirm Message

The "confirm" message is used to recover the route that was earlier declared as stale [9]. If the neighbor is sensed active again and there are some stale routes through it then there is a chance to recover those routes [9]. At this point the node sends the "confirm" message to the neighbor to ask whether the routes it held previously are active [9].

Route OK Message

This message is employed to respond to the "confirm" message. When a node receives a confirm message from one of its neighbors and it also has an active route to this desired destination then it replies with a "ROUTE OK" message to its neighbor [9].

Link Broken Message

The stale routes information is kept in the cache for some time and when it expires and receives no matching "ROUTE OK" message then the route will be removed from the cache [9]. When a node gets rid of the route it also broadcasts a "link broken" message to its neighbors [9]. All the neighbors then look up in their cache and eliminate that route from their cache if found [9].

The Protocol Description

The following is the pseudo-code version of the protocol for signal strength determination [9].

```
If (SS>=Threshold)
{ If (Packet sender does not belong to its neighbors)
{ Add it into the neighbors.
If (A stale route belongs to packet sender)
{Send confirm message to it.}
}
}
Else If (SS< Threshold)
{ If (Packet sender is one of the neighbors)
{ Discard it from the neighbors.
If (A route with packet sender as next node)
{ Mark the route as stale.
Record the marking time.
}
}
}
```

5.7.1.2 Advantages and Disadvantages

The advantages of the protocol involve the following: (1) It is compatible with DSR and no overhead is incurred [9]. (2) It only keeps an eye on the neighbor links' stability and exchanges route information with neighbor nodes. The local mechanisms will not bring about too much overhead [9]. (3) It is fault tolerant. Link failure can be detected in advance to maintain the returned route correctly and recover the cache when a host fails [9].

The drawback of this protocol lies in that nodes have to passively wait for packets, measure the signal strength, and use timers to refresh when there are no packets [9].

5.7.2 Survivable Routing Protocols in Ad-Hoc Wireless Networks

5.7.2.1 Introduction

One of the major concerns in mobile ad-hoc networks lies in the survivability of routing protocols facing attacks and failures. In the paper in [10], the authors have proposed the "on-demand secure routing" protocol to counter Byzantine attacks in ad-hoc networks and also explain how the protocol functions to survive.

5.7.2.2 Byzantine Attacks

Mobile ad-hoc networks are prone to a number of Byzantine attacks. Four main Byzantine attacks include "Black Holes," "Flood Rushing," "Wormholes," and "Overlay Network Wormholes" [10]. In a Black Hole attack the faulty nodes may drop entire or parts of data packets [10]. The adversary takes control of nodes in the network and whenever a node is chosen in a route data will be missing entirely or partially in that path [10].

In the case of an attack of Flood Rushing, the adversary utilizes techniques of suppressing flood, which has been applied by a few protocols [10]. Suppose that an attacker floods a network with an unauthenticated packet before the authentic version through the legitimate path. Then the unauthenticated packet will be propagated while the legitimate version may be ignored [10]. Adversaries can have access to authenticated nodes' Authentication, so techniques cannot prevent the attack effectively.

In a "Byzantine Wormhole" attack, two conspiring adversaries work together to send packets between one another so as to create a wormhole in the network [10]. Usually, the tunnel between the adversaries can be set up through a private channel. The adversaries can play tricks to be chosen with high probability by appearing with low cost [10]. Then they can easily attack the network by dropping packets.

In the attack of "Byzantine Overlay Network Wormhole," a number of wormholes are present in the network and they cooperate and form their own overlay network [10]. By exchanging packets between them, they make the routing protocol believe that they are all neighbors [10].

5.7.2.3 On-Demand Source Routing Protocol

On-demand source routing protocol (ODSBR), proposed in [10], can withstand a wide variety of Byzantine attacks. The ODSBR copes with attacks or failures using a consistent framework [10]. A fault can refer to any failure that may cause great loss or delay. It can result from external adversaries, Byzantine behavior, or merely by bursting traffic [10]. Under the scheme of ODSBR, the protocol finds a route without adversaries between source and destination, if any, with the assumption that only the source and destination nodes are trustworthy [10].

The protocol performs in three phases: least weight route discovery, Byzantine fault localization, and link weight management [10]. To discover a route, reliability matrix is exploited, which is comprised of a list of link weights where higher weight means lower reliability [10]. The matrix is updated at each node whenever a fault link is detected. Adaptive probing scheme is utilized to identify faulty links in the fault-finding phase, which are then gotten rid of when choosing a new route [10].

Route Discovery

In this phase, the lowest cost path can always be found with the help of "double flooding" and "per node flood verification" techniques [10]. However, the shortest path found is not necessarily fault free because adversary links may be chosen due to their low cost [10].

Fault Localization

Intermediate nodes involved in the route are supposed to reply to the source with secure acknowledgments, which can serve as the evidence that packets have been sent to the destination with success and without corruption [10]. Because of the scheme of probing, a malicious node is only able to commit a fault in one of its adjacent links [10].

Link Weight Management

The adversary link weight is increased and kept until a required number of reliable acknowledgments have been received through it [10]. It is efficient to limit the general loss rate even when dynamic adversaries are present that can adjust its role from good to bad or vice versa [10].

Provided that there is a route without fault nodes from the source to the destination, the above three phases can help to reduce the packet loss rate caused by faulty nodes, even in the presence of more than half nodes that collude with adversaries committing Byzantine faults [10].

5.7.2.4 Attacks Mitigation

The protocol ODSBR proposed by [10] can resist all the following attacks.

Flood Rushing

The ODSBR protocol makes it possible to check the authentication and integrity of the packets used for "route discovery flood" hop by hop, which is effective to prevent invalid flood packets from blocking the valid packets or propagating through the network [10].

Black Hole

The ODSBR protocol employs point-to-point acknowledgments to find the black hole attack [10]. Once it detects an attack, it tries to discover the location of the attack [10]. Through probing process, the spot of adversary can be limited to one link and its weight will be doubled, which leads the protocol not to select the path having that link [10]. ODSBR can always find a path with no adversary to a destination as long as one exists.

Byzantine Wormhole

ODSBR looks at Byzantine wormhole attacks from a different perspective. This wormhole attack will appear as a faulty link to this protocol rather than a wormhole [10]. ODSBR withstands this attack by detecting its existence by increasing the weight of the link with the wormhole rather than preventing it from happening [10]. The packet loss rate and the time spent to find the path without any adversary will increase proportionally along with the number of wormhole links present in the network [10].

Byzantine Overlay Network of Wormholes

If the number of adversaries in a network is very large, then the convergence of ODSBR is slowed down but it can bound the attacks created by the adversaries [10].

5.7.2.5 Experimental Results

The protocol is implemented and several simulations of these attacks are done. It has been proven that the ODSBR can withstand all of these attacks efficiently [10]. The protocol performs with decreasing efficiency only when it has to find and try to avoid many faulty links [10].

5.7.3 Fault-Tolerance Techniques for Wireless Ad-Hoc Sensor Networks

Each node in a wireless ad-hoc sensor network consists of five components: "computation, communication, storage, sensors, and actuators" [11]. To construct an efficient embedded sensor network, a few factors have to be taken into account, such as cost, overhead, and fault tolerance. The authors of [11] proposed a cost-effective solution for fault tolerance in sensor networks, and they also developed techniques and algorithms to efficiently overcome the faults and make the wireless network fault-free.

The technique proposed is heterogeneous fault tolerance, where several different kinds of resources are backed up with a single kind of resource [11]. According to the paper, all of the five kinds of resources, "computation, communication, storage, sensors, and actuation," in wireless sensor networks can replace each other with suitable changes in system and application software. Because the sensing part is considered to have the highest fault rates and the multimodal sensor fusion is essential in the successful application of a wireless ad-hoc network, they focus mainly on providing backups for one type of sensor with another [11]. The mechanism of designing a heterogeneous backup with low overhead will be explained in detail as follows.

The authors of [11] have proposed two fault models to deal with the faults in sensors in wireless sensor networks. The first model applies to the sensors producing binary outputs [11]. This model has a straightforward fault detection procedure that just studies the sensor's output and determines whether the sensor is functional rather than capturing the probability and statistics of erroneous reported results [11]. The second fault model is associated with sensors having continuous and multilevel digital outputs. The advantages of this approach are that it can be utilized for many fault models and is quite useful in case of transient errors and errors in measurements, and it also addresses fault detection and correction at the same place [11].

5.7.4 Fault-Tolerance Issues in Wireless Sensor Networks

5.7.4.1 Introduction

During the past several years, fault-tolerance issues in sensor networks have drawn much attention from researchers. One major problem with sensors is that their energy is limited and hence they cannot support long-haul communication; therefore, they need architecture with multiple layers to transmit the data. A highly efficient scheme proposed in [4] is to divide the network into different clusters where each cluster has a node as the cluster head called gateway with sufficient energy. Cluster-based gateway switch routing protocol organizes nodes into clusters. Every cluster elects a node to act as the cluster head.

Cluster heads, special nodes responsible for the routing process, can communicate with each other by using gateway nodes. A gateway node is defined as a node that has two or more cluster heads as its neighbors [4]. In a sensor network if a cluster head is damaged, all the nodes in the cluster are no longer accessible [4].

5.7.4.2 Sensor Network Clustering

Under the scheme proposed by the paper in [4], gateway nodes are those with less energy constraint and group sensor nodes are in different clusters in the wireless network. Except for the energy constraint, a gateway has the same kind of properties as an ordinary sensor. Clusters are formed by evaluating the communication distance between gateways and sensors and the load on gateways. One node in a cluster communicates with nodes in another cluster only through the gateway of the cluster.

5.7.4.3 Fault-Tolerance Mechanism for Clustered Sensor Network

To ensure the reliability of a sensor network, the system is supposed to perform runtime recovery from the failure of a gateway in a cluster. A mechanism is proposed by the paper, which can be interpreted in two steps: "detection and recovery" [4]. In the paper [4], the authors assume that the sensor network is stationary and nodes, including gateways and sensors, communicate over a single shared channel. Also, each node can be aware of its location with the assistance from an external orientation system. Gateways are all presumed to be in each other's communication range and can communicate status information.

5.7.4.4 Gateway Failure Detection

Failure detection is indispensable as the first step in the system of fault tolerance. Gateways are independent and responsible for sensors in their own clusters. Periodic status updates are done through inter-gateway communication. Gateways come to know about clusters in the system through status updates. The scheme of periodic status updates via inter-gateway communication is exploited to allow gateways to exchange their status with the rest of clusters [4].

The TDMA MAC protocol is used for communication in implementation [4]. Sensors use slots allocated by gateways to send data (Figure 5.3 shows slot allocation). During these slots, a sensor is scheduled to send data to the gateway in the cluster according to the information like energy, tasks, and priority [4]. During "Route Update" slot, the gateway informs sensors of the information on scheduling and routing. The slots of route update are the dark slots, and

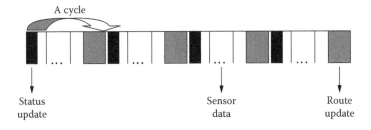

Figure 5.3   Slot allocation in sensor networks. (After G. Gupta and M. Younis, *Proc. of IEEE Wireless Communications and Networking Conference,* 3: 1579-1584, 2003.)

white slots indicate the slot for sensors to send data and energy status in that life cycle [4].

One cycle is considered completed after all sensors in a cluster have sent energy and data status to the gateway. As a cycle ends, a "Status" is created by the gateway, which includes information on the status of both itself and sensors in the cluster [4].

In the "Status Update" slot, gateway status is exchanged between gateways through inter-gateway communication [4]. A "Multiplicative Increase Linear Decrease" (MILD) scheme is used for scheduling status exchange [4]. The algorithm increases the time period for an exchange by a "multiplicative factor" when faults are absent, while time period is linearly decreased in the presence of fault [4]. This method helps in reducing the overhead caused by exchanging status when the system is reliable and recovering quickly from failures [4].

To deal with link failures, a simple forwarding approach is adopted. Any new update received by one gateway is forwarded to all other gateways that are in its transmission range [4]. Redundant messages are added in the network by this method when there is no fault in the network, but it only guarantees that each gateway maintains the same information about status [4]. Updates of a gateway cannot be received by other gateways if the gateway fails. In this case, recovery will be started [4].

In Figure 5.4 all gateways A, B, C, D are in direct communication range and hence they form a Fully Connected Gateway Model. Table 5.1 shows an example of a fully connected network with no forwarding of update messages [4].

Figure 5.5 shows link failure between gateways A and C and also complete failure of gateway D [4]. Shown in Table 5.2, gateway B tries to analyze multiple link failure by taking into account that none of the status update message is received from gateway D. Also, D has not sent experiences to any of the other gateways. Therefore, other gateways come to notice that D cannot send data to other nodes

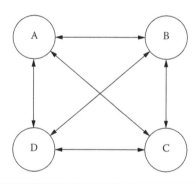

Figure 5.4 Gateway model that is fully connected. (After G. Gupta and M. Younis, *Proc. of IEEE Wireless Communications and Networking Conference,* **3: 1579-1584, 2003.)**

Table 5.1   The Table for a Fully Connected Network

	A	*B*	*C*	*D*
A	*	1	1	1
B	1	*	1	1
C	1	1	*	1
D	1	1	1	*

*   Denotes own update, 1 shows that the update is received, 0 represents the update is lost.

Source: After G. Gupta and M. Younis, Proc. of IEEE Wireless Communications and Networking Conference, 3: 1579-1584, 2003.

because of transmitter failure [4]. Then D is considered to fail and sensors in cluster of D are supposed to be recovered. [4].

5.7.4.5 Recovery

In the recovery phase, the type of faults is identified first and then the sensors are allocated to the new clusters. Parsing of status information is performed to know the sensors' identity that have lost communication with the failed gateway. During clustering, a range set is created by each gateway based on the distance from those sensors. A sensor is recovered by being assigned to another gateway if the sensor appears in the range set of the gateway. Assuming that a sensor is included in more than one range set of other gateways, the sensor will be

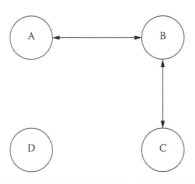

Figure 5.5   Gateway model of multiple links with one complete failure. (After G. Gupta and M. Younis, *Proc. of IEEE Wireless Communications and Networking Conference,* 3: 1579-1584, 2003.)

Table 5.2   The Table for a Network with One Complete Failure

	A	B	C	D
A	*	1	0	0
B	1	*	1	0
C	0	1	*	0
D	0	0	0	0

* Denotes own update, 1 shows that the update is received, 0 represents the update is lost.

Source: After G. Gupta and M. Younis, Proc. of IEEE Wireless Communications and Networking Conference, 3: 1579-1584, 2003.

accepted by the gateway that possesses the lowest cost of communicating with the sensor.

5.7.5 Ad-Hoc Routing Service That Is Fault Tolerant under Adversarial Circumstances

In a mobile ad-hoc network with adversarial nodes, routing performance may be greatly degraded. Most existing approaches to alleviating the damage caused by malicious attacks fall into two categories: one is the prevention of attacks beforehand in which extra protection is imposed on the router and also setting certain rules for cooperation in routing, and the other is detection in which all the adversarial nodes are first detected and marked, and then they are eliminated while considering a route from source to destination. But this technique is difficult and expensive [15]. The paper in [15] proposes a new routing algorithm known as "Best-Effort Fault Tolerant Routing" (BFTR), which uses network redundancy to find a non-adversarial path from source to destination. The goal of the algorithm is to offer data routing service and at the same time to gain higher delivery ratio with lower overhead even in the face of faulty nodes [15].

5.7.5.1 Design Philosophy

The BFTR utilizes the network's redundancy, which is the intrinsic nature of an ad-hoc network, for providing high delivery ratio even in the face of malicious nodes [15]. Consider the networks shown in Figure 5.6 from the paper in [15].

Consider the two cases shown in Figure 5.6: in (a) there is no good path between source and destination so this problem cannot be solved by any routing protocol, whereas in (b) there is one good path and one bad path between source S and destination D [15]. But most of the routing protocols such as DSR are not able to find a good

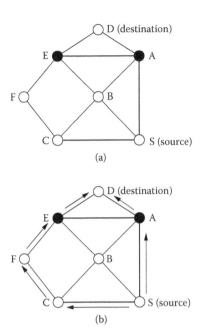

Figure 5.6 **(a) There is no good route between source S and destination D; (b) the shortest route S→A→D is bad, but S→C→F→E→D is a good route.**

path between source and destination because of their inability to find all the paths between source and destination. When DSR floods the route request and a node gets that request for the second time it just drops that packet [15]. DSR does not care for the issue of packet delivery and that's why in the above example DSR has high probability to discover a route that includes malicious nodes [15]. Instead of excluding or deleting misbehaving nodes, BFTR will tolerate them to a certain extent and thus can maintain high delivery ratio even in the face of faulty nodes [15].

BFTR takes into account the pattern in which good nodes send messages to each other. From the end-to-end perspective, a good path shows the same pattern of behavior [15]. Those paths that deviate from such a pattern are considered bad paths (paths having misbehaving nodes). The basic features of BFTR are given below [15].

BFTR requires the existence of prior trust relationships between the sender and the receiver [15]. The source can determine whether a packet has been sent with success by evaluating the end-to-end execution, so it doesn't need any security assistance from other neighboring nodes [15].

The BFTR does not differentiate among attacks, so that it is effective for all kinds of attacks [15]. The success of individual packet delivery is exploited to evaluate the performance of the routing service [15].

5.7.5.2 Best-Effort Fault Tolerant Routing Algorithm

The algorithm of BFTR is based on a couple of assumptions. One is that the sender and the receiver are assumed to be well-behaved nodes and there is no routing service for misbehaving nodes [15]. Another assumption is that there should be some authentication process between source and destination. The authentication can be achieved through a public key infrastructure, or the sender and receiver can negotiate on a secret key through some secure channels.

BFTR is basically a "source routing algorithm." It employs flooding for retrieving a set of routes between the sender and the receiver, which is similar to the DSR. BFTR then chooses the shortest path from a set of paths available. A path is identified as a valid one if for a certain observation window all the packets have been delivered with high packet delivery ratio [15]. Then BFTR chooses the shortest feasible path available to route the packets between source and destination. For practical path modeling, the random variable $Y(\pi)$ is defined as follows given a path π [15]: $Y(\pi)$ is 1 if π successfully sends the observed packet with the delay smaller or equal to D, otherwise it equals 0.

The probability that a packet is delivered successfully is given by $(Pi\{Y(\pi) = 1\})$, denoted as $p(\pi)$, satisfying $p(\pi) \geq p_0$ [15]. p_0 is the element of interval $(0,1)$ and is defined as the expected packet delivery ratio [15]. Then we can determine whether a particular path (π) is feasible or not [15]: π is a feasible path if $p(\pi) = p_0$; π is an infeasible path if $p(\pi) < p_0$.

The fundamental difference between BFTR and DSR is that BFTR requires that the nodes along the route send route reply packets to the source on the reversing path of the route request packet, while DSR requires that the destination send several replies to the source so that the source will have several paths to send packets to the destination [15]. The route reply packets in BFTR are encrypted with the help of a secret key shared by both the source and destination [15]. The aim is to prevent fabrication of messages and save them from replay attacks [15]. Now from these paths the sender node makes its decision to choose one using BFTR protocol [15]. For determining the feasibility of a route, the source should know how many packets have been received along the route under consideration [15]. The destination in this case sends a feedback to the source about the number of packets received along the path by piggybacking onto acknowledgments [15]. A route is considered a feasible route if the sender is able to send packets and receive acknowledgments with a delay of less than or equal to 2*D where D is considered as the one-way delay. If the acknowledgment is not received within this time then the packet is considered to be lost and the route is marked as a malicious route [15].

By this means, BFTR is able to find appropriate routes by evaluating the packet delay of one round trip [15]. However, if the delay is large, the path may not be selected, because a large delay shows that the network condition on the route is not good. The value of delay D is crucial to the BFTR algorithm for the reason that if the value of D is small then BFTR may wrongly reject the feasible paths, whereas the value of D being too large may result in acceptance of infeasible paths [15]. For simulation experiments, the value of delay is set equal to the expected round trip time.

BFTR route maintenance is the same as that of DSR. For a particular report of route failure, the current route will be abandoned and another shortest route in the route cache will be chosen to replace the invalid one. If in the path cache all routes have been declared as infeasible then BFTR will initiate a new route discovery [15].

BFTR does not ensure the performance of routing. The packet delivery loss will increase with the growth in the number of misbehaving nodes [15]. But performance can also be lowered in the presence of network congestion. BFTR does not distinguish between these two cases [15].

5.7.6 Fault-Tolerant Broadcast Delivery in Mobile Ad-Hoc Networks

The paper in [13] proposes a protocol that provides a fault-tolerant and dependable broadcast. The protocol includes two phases: in phase one a message packet is scattered to all the receivers, and in phase two all acknowledgment messages are received from all the receivers. In the scattering phase, a source-based forwarding tree will be built up [13]. Because the nodes are mobile, there is a possibility of change in the topology [13]. If the change rate is very high and the network cannot maintain the forwarding tree, the protocol will change to flooding. The system proposed is composed of say n mobile nodes that communicate via packet radio network [13]. For all hosts in the system, a clustering protocol is employed to build up a clustered architecture. A cluster head is selected to coordinate the communication between nodes in the cluster and inter-clusters [13].

5.7.6.1 Definitions and Assumptions

The model proposed has certain assumptions listed below [13]:

> Definition of *Reliable Broadcast*: Each host in a network can broadcast a message, which is to be delivered to all the active nodes so that it meets the following requirements: validity, integrity, and termination [13].
>
> Definition of *Eventual Subsidence*: The resources allocated for each broadcast message m, like storage and bandwidth, are eventually released when all of the acknowledgments for that message are being received [13]. In order to prevent the protocol from running indefinitely due to faulty conditions and host mobility, the technique requires two properties that must be satisfied [13].
>
> *Property of Liveliness*: If a node is linked to a cluster head, it remains linked before the communication of status information between the node and the cluster head ends [13].

Property of *Cluster Head Stability*: During the time required to exchange the information with the other cluster heads, the host itself acts as cluster head, if there are partially diffused messages [13].

5.7.6.2 The Reliable Broadcast Protocol

A unique pair of attributes is assigned to a particular broadcast message <sender ID, sender sequence number>. A message can be regarded as stable only when all its destinations have received it [13]. The bandwidth and host memory for a broadcast message are released as soon as the message becomes stable [13].

5.7.6.3 The Forwarding Tree

During the scattering phase, in order to scatter the broadcast message to all the destinations, a routing tree with the source in the center is used on-demand. This tree is known as a forwarding tree (FT) and it requires no route maintenance in spite of mobility. The tree is constructed by connecting adjacent cluster heads with the help of common gateways [13]. The lifetime of this tree is until an acknowledgment message is received after the scattering phase. The FT, instead of following the shortest path, follows the fastest path for message scattering between source and destination. If the FT is broken due to mobility of nodes and if the cluster head has no way to route acknowledgment messages, there is no packet recovery and the protocol resorts to flooding [13].

5.7.6.4 Message Stability

The reliable broadcast protocol proposed adopts a packet retransmission policy to deal with the packet loss caused by link failures [13]. The time interval during which a source node waits for an acknowledgment from the recipient of the message is similar to the delay period of two round trips [13]. The cluster heads in the FT (Forwarding Tree) perform packet recovery where is in their charge. One salient feature of the approach is that it causes packet recovery to be localized in the area of the failure and thus avoids the start of retransmission from the originator's cluster head, despite resulting in extra occupation of memory in each node for the reason that the host cannot discard a message until it is sure that the message is stable [13].

5.7.6.5 Topology Change and Host Mobility

The protocol proposed works well even if there is a change in the network topology. Suppose a host *p* moves from cluster *A* to cluster *B*. Two cases need to be considered: (1) After *p* sends a message, *p* moves while the message is unstable, and (2) *p* is the message's destination and the cluster head of *A* has not received acknowledgment from p [13]. According to the proposed protocol, both cases can be solved if p exchanges status messages with B. The status messages include their respective stable and unstable information, missing messages, and updated acknowledge status information [13]. Consider another case where cluster head changes and the old cluster head still has some messages in the following status: (1) diffusion of those messages have not been initiated, or (2) diffusion has only been partially done, or (3) the old cluster head has learned the stable state of some messages but it has not sent the information out [13]. For (1), they argue that the sender must be forced to restart the broadcast of m, as the rest of the cluster heads don't have the message. In cases (2) and (3) the new cluster head must get the status information of the old cluster head. This can be achieved by all the members of the cluster sending status to the new cluster head. If the old cluster head has joined a new cluster head then it can pass on its status to the current cluster head to inform the current cluster head that it has been a cluster head [13].

The new cluster head gets the local status from cluster nodes by exchanging status information. Then it passes this information to all other cluster heads. Similarly other cluster heads reply with their status information possibly merged with information from an old cluster head that appears in the cluster [13].

5.7.6.6 Experimental Results

They are currently in the process of running simulations, and their first results are quite encouraging. When cluster heads move slowly, congestion is the main cause of packet loss. They are also working on a "link state routing protocol" that exploits the existing reliable broadcast protocols instead of flooding [13].

5.8 Fault-Tolerant Distributed Consensus

5.8.1 Consensus Problem Overview

A fault-tolerant system is supposed to withstand a certain number of failures of communication or process, no matter whether the failures are intermittent or permanent [38].

It is critical for any fault-tolerant system to deal properly with the consensus problem as correct processes can only make an agreement through consensus on whether a fault has happened and how to avoid it [38]. To launch a consensus, each process brings up a value respectively to form a value set, from which one value should be agreed on by all correct processes [38].

It is commonly considered to be difficult to solve consensus problems in a distributed system in which processes may fail. Different schemes and algorithms are required by systems with different sorts of failures and variable levels of synchrony. In a completely synchronous system, the upper bounds are set to the communication and processing delay. However, in a totally asynchronous system, no such bounds are defined to regulate the relative speed of the communicating and processing. There are two types of failures involved: Byzantine failures in which a malicious process may send incorrect data, and Fail-Stop failures that may kill a process arbitrarily and prevent it from continuing to participate in the algorithm.

5.8.2 Consensus in Synchronous Systems

5.8.2.1 Synchronous System

A synchronous system can be described with a model that is composed of a finite set of processes, namely, $S = \{p_1,...,p_n\}$ [39]. Those processes synchronize and communicate by exchanging packets along channels [39]. The channel that links processes p_i and p_j, is denoted as (p_1, p_n) [39].

The system can be regarded as synchronous when each of its executions contains a series of rounds indicated by consecutive integers 1, 2, 3, etc. A round includes three successive phases [39]:

The sending phase in which processes send messages.
The receiving phase during which processes receive messages.
The execution phase in which processes perform local computation after receiving messages in that round.

The crucial characteristic of the model involves a message being sent and received by different processes at the identical round [39]. The model is also based on the assumption that the communication system is failure-free [39].

5.8.2.2 A Model for Process Failures

A process will be suspected to be faulty if it acts differently from the rules formulated in the algorithm. Categorizing different failure types is necessary to the research. In the paper in [39], three failure models are considered:

Byzantine Failure: A faulty process may act arbitrarily.
Crash Failure: A misbehaving process may stop arbitrarily during its performance. After the crash, the process is permanently dead.
Omission Failure: A misbehaving process may omit to send or receive messages it is expected to.

Actually, the severity of the three failure models is various in the following sequence: Byzantine > Omission > Crash. If a protocol intends to deal with problems in the failure models with high severity, it has to solve the same problem in the failure models with comparatively low severity [39].

5.8.2.3 Consensus Problems

The authors of [39] have proposed three concepts to define a consensus problem:

Termination: Each and every correct process will finally make a decision.

Validity: If a process agrees on a value then the value must be one from some process.

Agreement: The final value decided should be agreed on by all correct processes.

Notice that according to the agreement property a misbehaving process is allowed to propose a different value from the one agreed on by all correct processes. Based on the point of weakness, the uniform agreement property is proposed and defined as:

Uniform Agreement: Every two processes, no matter correctness or fault, must decide the same value.

5.8.3 Consensus with Probability

It seems that real consensus in an asynchronous distributed system is impossible. However, we can make it possible by redefining "consensus." Rather than being critical to require consensus in a finite period of time, we only ask for consensus with certain probability.

In the paper in [18], the authors have proposed an algorithm to resist both Byzantine failures and fail-stop failures. The algorithm is successful to achieve consensus with probability if $n > 2m$ in the fail-stop failure model and $n > 3m$ in the Byzantine failure model, where n denotes the total number of nodes and m the number of faulty nodes [18].

Under the fail-stop tolerant scheme, each process expects only $n - m$ acknowledgments from other processes after broadcasting a value [18]. After having received $n-m$ acknowledgments, a decision has to be made according to $n - m$ responses. The problem becomes that whether a decision made based on the $n - m$ responses is identical to that made by other processes [18]. Actually, those $n - m$ process that send back acknowledgments may be different [18].

The algorithm can be described in several phases. First, in each round every process broadcasts a favorite value together with the number of processes indicated by S, which are in favor of the same value and expect $n - m$ answers [18]. Then each process will change its favorite value to the one that is favored by most processes [18]. It goes on in this way until one process receives m different messages of a same value, each of which is favored by at least $n/2$ processes [18]. The algorithm will stop before deciding on the value in the following two phases [18].

Although consensus with certain probability may not meet some strict requirements, it can be practical in several real systems.

5.9 Conclusions

The demands of fault-tolerant systems are expected to increase dramatically in the near future. When we become more and more dependent on Internet and computing services, we have to rely on these fault-tolerant systems to minimize the impact of loss because of the fact that the failure behaviors are unpredictable in the Internet. Although problems and faults can happen in ad-hoc networks, the fault-tolerant techniques to deal with them are abundant. The algorithms that we can use depend greatly on the properties of distributed systems, the assumptions and trade-offs that we make, and the kinds of faults that we expect.

Acknowledgments

This work is supported in part by the U.S. National Science Foundation (NSF) under the grant numbers CCF-0829827, CNS-0716211, and CNS-0737325.

References

1. H. Kawahigashi, Y. Terashima, N. Miyauchi, and T. Nakakawaji, "Designing Fault Tolerant Ad Hoc Networks," *Proc. of IEEE Military Communications Conference*, 3: 1360-1367, 2005.
2. P. Basu, W. Ke, and T.D.C. Little, "Dynamic Task-Based Anycasting in Mobile Ad Hoc Networks," *Mobile Networks and Applications*, 8(5), 593-612, 2003.
3. M. Castro and B. Liskov, "Practical Byzantine Fault Tolerance," *Proc. of the 3rd Symposium on Operating Systems Design and Implementation*, 173-186, 1999.
4. G. Gupta and M. Younis, "Fault-Tolerant Clustering of Wireless Sensor Networks," *Proc. of IEEE Wireless Communications and Networking Conference*, 3: 1579-1584, 2003.
5. H.F. Lipson and D.A. Fisher, "Survivability—A New Technical and Business Perspective on Security," *Proc. of the ACM 1999 New Security Paradigms Workshop*, 1-7, 1999.
6. D. Chen, S. Garg, and K.S. Trivedi, "Network Survivability Performance Evaluation: A Quantitative Approach with Applications in Wireless Ad-Hoc Networks," *Proc. of the 5th ACM International Workshop on Modeling Analysis and Simulation of Wireless and Mobile Systems*, 61-68, 2002.
7. M. Virendra, S. Upadhyaya, V. Kumar, and V. Anand, "SAWAN: A Survivable Architecture for Wireless LANs," *Proc. of the 3rd IEEE International Workshop on Information Assurance*, 71-82, 2005.
8. J. McDermott, A. Kim, and J. Froscher, "Merging Paradigms of Survivability and Security: Stochastic Faults and Designed Faults," *Proc. of the 2003 Workshop on New Security Paradigms*, 19-25, 2003.
9. C.-H. Chuan and S.-Y. Kuo, "Cache Management of Dynamic Source Routing for Fault Tolerance in Mobile Ad Hoc Networks," *Proc. of Pacific Rim International Symposium on Dependable Computing*, 199-205, 2001.

10. B. Awerbuch, R. Curtmola, D. Holmer, H. Rubens, and C. Nita-Rotaru, "On the Survivability of Routing Protocols in Ad Hoc Wireless Networks," *First International Conference on Security and Privacy for Emerging Areas in Communications Networks*, 327-338, 2005.

11. F. Koushanfar, M. Potkonjak, and A. Sangiovanni-Vincentell, "Fault Tolerance Techniques for Wireless Ad Hoc Sensor Networks," *Proc. of IEEE Sensors*, 1491-1496, 2002.

12. Y.-A. Huang and W. Lee, "A Cooperative Intrusion Detection System for Ad Hoc Networks," *Proc. of the 1st ACM Workshop on Security of Ad Hoc and Sensor Networks*, 135-147, 2003.

13. E. Pagani and G.P. Rossi, "Providing Reliable and Fault Tolerant Broadcast Delivery in Mobile Ad-Hoc Networks," *Mobile Networks and Applications*, 4(3), 175-192, 1999.

14. H.-W. Tsai, T.-S. Chen, and C.-P. Chu, "An On-Demand Routing Protocol with Backtracking for Mobile Ad Hoc Networks," *IEEE Wireless Communications and Networking Conference*, 3: 1557-1562, 2004.

15. Y. Xue and K. Nahrstedt, "Providing Fault-Tolerant Ad Hoc Routing Service in Adversarial Environments," *Wireless Personal Communications*, 29: 367-388, 2004.

16. M. Ben-Or, "Another Advantage of Free Choice: Completely Asynchronous Agreement Protocols," *ACM Symposium on Principles of Distributed Computing*, 27-30, 1983.

17. G. Bracha, "An Expected Rounds Randomized Byzantine Generals Protocol," *J. ACM*, 34(4), 910-920, 1987.

18. G. Bracha and S. Toueg, "Asynchronous Consensus and Broadcast Protocols," *J. ACM*, 32(4), 824-840, 1985.

19. T.D. Chandra, V. Hadzilacos, and S. Toueg, "The Weakest Failure Detector for Solving Consensus," *ACM Symposium on Principles of Distributed Computing*, 147-158, 1992.

20. T.D. Chandra and S. Toueg, "Unreliable Failure Detectors for Asynchronous Systems," *ACM Symposium on Principles of Distributed Computing*, 325-340, 1991.

21. D. Dolev, N.A. Lynch, S.S. Pinter, E.W. Stark, and W.E. Weihl, "Reaching Approximate Agreement in the Presence of Faults," *J. ACM*, 33(3), 499-516, 1986.

22. D. Dolev, R. Ruediger, and H.R. Strong, "Early Stopping in Byzantine Agreement," *J. ACM*, 37(4), 720-741, 1990.

23. M.J. Ficher, N.A. Lynch, and M.S. Paterson, "Impossibility of Distributed Consensus with One Faulty Process," *J. ACM*, 32(2), 374-382, 1985.

24. V. Hadzilacos and J.Y. Halpern, "Message-Optimal Protocols for Byzantine Agreement," *ACM Symposium on Principles of Distributed Computing*, 309-323, 1991.

25. J.Y. Halpern, Y. Moses, and O. Waarts, "A Characterization of Eventual Byzantine Agreement," *ACM Symposium on Principles of Distributed Computing*, 333-346, 1990.

26. T. Chandra and S. Toueg, "Unreliable Failure Detectors for Reliable Distributed Systems," *J. ACM*, 225-267, 1996.

27. C. Dwork, N. Lynch, and L. Stockmeyer, "Consensus in the Presence of Partial Synchrony," *J. ACM*, 35(2), 288-323, 1988.

28. P. Verissimo and C. Almeida, "Quasi Synchronism: A Step Away from the Traditional Fault-Tolerant Real-Time System Models," *IEEE TCOS Bulletin*, 7(4), 35-39, 1995.

29. F. Cristian and C. Fetzer, "The Timed Asynchronous Distributed System Model," *Proc. of 28th Int. Symp. On Fault-Tolerant Computing*, 140-149, 1998.

30. G. Neiger and S. Toueg, "Automatically Increasing the Fault-Tolerance of Distributed Algorithms," *Journal of Algorithms*, 11(3), 374-419, 1990.

31. M. Malek, "Omniscence, Consensus, Autonomy: Three Tempting Roads to Responsiveness," *Proc. of 14th IEEE Symposium on Reliable Distributed Systems,* 12-14, 1995.

32. F. Cristian, B. Dancey, and J. Dehn, "Fault-Tolerance in the Advanced Automation System," *Proc. of the 4th Workshop on ACM SIGOPS Enropean Workshop,* 6-17, 1990.

33. M. Barkorak, A. Dahbura, and M. Malek, "The Consensus Problem in Fault-Tolerant Computing," *ACM Computing Surveys,* 25(2), 171-220, 1993.

34. D. Dolev, C. Dwork, and L. Stockmeyer, "On the Minimal Synchronism Needed for Distributed Consensus," *J. ACM,* 34(1), 77-97, 1987.

35. L. Lamport, R. Shostak, and M. Pease, "The Byzantine Generals Problem," *ACM Transactions on Programming Languages and Systems,* 4(3), 382-401, 1982.

36. R.L. Rivest, A. Shamir, and L. Adleman, "A Method for Obtaining Digital Signatures and Public-key Cryptosystems," *Communications of the ACM,* 21(2), 120-126, 1978.

37. J. Turek and D. Shasha, "The Many Faces of Consensus in Distributed Systems," *IEEE Computer,* 25(6), 8-17, June 1992.

38. L. Kesteloot, Fault-Tolerant Distributed Consensus. January 20, 1995. http://www.teamten.com/lawrence/papers.html

39. M. Raynal, "Consensus in Synchronous Systems: A Concise Guided Tour," *9th IEEE Pacific Rim Int. Symposium on Dependable Computing (PRDC'2002),* IEEE Computer Society Press, 221-228, Tsukuba (Japan), December 2002.

40. Kazakos, D. "A Generalized Model for Network Survivability," *Proceedings of the 2003 Conference on Diversity in Computing—TAPIA 03,* 2003.

41. D. Yang. "A Quantitative Survivability Evaluation Model for Wireless Sensor Networks," *2006 IEEE International Conference on Networking Sensing and Control,* 2006.

42. V. Anand, "SAWAN: A Survivable Architecture for Wireless LANs," *Third IEEE International Workshop on Information Assurance (IWIA 05),* 2005.

43. M. Guida, "Reliability and Survivability Methodologies for Next Generation Networks," *Proceedings of the 6th International Conference on Advances in Mobile Computing and Multimedia—08,* 2008.

44. K.S. Trivedi, "Network Survivability Performance Evaluation," *Proceedings of the 5th ACM International Workshop on Modeling Analysis and Simulation of Wireless and Mobile systems—MSWiM 02,* 2002.

45. P. Basu, K. Wang, and T.D.C. Little, "Dynamic Task Based Anycasting in Mobile Ad Hoc Networks," *MCL Technical Report No. 07-22-2003.*

Chapter 6

Time Synchronization in Sensor Networks and Underwater Sensor Networks

Michael Galloway, Yanping Zhang,
Yang Xiao, and Peng Shao

Contents

This chapter presents methods for synchronization of time and data in sensor networks and underwater sensor networks. The ability to synchronize nodes in a wireless network is a very important factor for energy efficiency and successful data transfer. Synchronizing each node will help maintain unnecessary energy loss due to lost and retransmitted data. Several protocols, such as pair-wise and global time synchronizing, distributed and centralized time synchronizing, etc., are covered throughout this chapter. Security synchronization protocols for controlling attacks on wireless sensor networks are also covered.

6.1 Introduction

Wireless sensor networks are such a popular research area that they have a variety of applications, such as environmental observation, military monitoring, building monitoring, healthcare, and so on. They typically consist of a random number of nodes, each of which is equipped with special sensors for sensing interested phenomena. Together with sensors, nodes are capable of sensing, data processing, and communicating. Then they work together to process and route data they collect from the environment.

Clock synchronization is very important in wireless sensor networks [12–18, 20, 29, 30, 33]. It allows the nodes in the network to communicate successfully with minimal effort. Recent research leads to smaller sensors that consume lower power. Being small, a mote has a limited power supply. Energy efficiency is a major concern in implementing communication in wireless sensor networks. Also, a given sensor network could consist of a large number of nodes, so that protocol scalability is also an issue.

Generally, nodes route their data to a base station, which acts as a gateway to a wired back-end server. Otherwise, the nodes will collect data and store it until the lifespan of the network has been depleted. At this time, the nodes are collected and the data is removed by connecting it to an information retrieval unit in a laboratory. Sensor networks rely on the fact that the nodes, either current or new, must meet some constraints on their organization and operation. This chapter introduces some of the ways to meet those constraints. Clock synchronization in wireless sensor networks has more obstacles to overcome than that of wired networks [8]. Some of those obstacles are robustness, energy efficiency, and ad-hoc deployment. Robustness relates to the fact that there are no constant static connections between nodes. Nodes can be mobile, and therefore can change the topology of the network quickly. Also, nodes can fail due to depleting their batteries. The protocol used has to accommodate for this change. Energy efficiency is also a problem directly related to how communication and synchronization occurs within the network. This is stated by the fact nodes can only achieve global synchronization by sending messages to each other wirelessly [8].

Sensor nodes are always using oscillators to keep their local time. Furthermore, in order to reduce the cost of nodes, cheap oscillators are always used, which are instable and may change with temperature, atmospheric pressure, or voltage. This makes clock drift unavoidable. Then the nodes need to resynchronize their clocks every so often due to their local clock drift. It is also possible not to synchronize all of the nodes in the network at a set interval. Doing so would waste lots of energy. Some nodes can be permitted to ignore their clocks drift and resynchronize when necessary. In a typical wireless sensor network, nodes could be permitted to let their clocks go out of sync and resynchronize when needed. In this chapter, many clock synchronization protocols are covered.

We first give a total analysis of time synchronization, including its necessity in sensor networks, existing time synchronization protocols in wired networks, as well as the difference between wired and wireless networks, and the challenges and principals of its design in wireless sensor networks.

Time synchronization is crucial in handling energy efficiency in sensor networks. For the majority of the time, nodes in the network should be dormant to conserve energy [2]. The synchronization of each node's clock is important when the nodes awake to transmit data. If their clocks are not synchronized, their transmission time will seem random and the propagation of data through the network will be slow. Time-stamp synchronization will be discussed first. It is a widely used synchronization protocol applied in distributed sensor networks. A specific wireless sensor network protocol called TSync [2] will be discussed. This is a light-weight agile protocol suited for sensor networks.

Three global clock synchronization algorithms will be discussed. They are an all-node-based method, a cluster-based method, and a fully localized diffusion-based method [4]. The all-node-based method allows a single node in the network to initiate a global synchronization. The message is then transmitted to all other nodes in the network in a predetermined cycle. The cluster-based method works the same as the all-node-based method, but just with head cluster nodes. These head nodes synchronize with each other, then allow their child nodes to synchronize with them. The last method is the fully localized diffusion-based method. This method can run in synchronous or asynchronous mode. In synchronous mode, each node has to wait until every other node in the network has finished updating with its neighbor before starting another synchronization round. In the asynchronous mode, nodes in the network can initiate clock updates with their neighbors at any time they want. There is no time constraint.

The next topic covered is adaptive clock synchronization. The bounds for clock synchronization are given and different models are implemented. The next discussion relates synchronization protocols used by other types of networks and compares their protocols to ones that are needed in a wireless sensor network.

Delay measurement time synchronization is covered as a flexible energy-efficient protocol implemented on Berkley motes [6]. This protocol is based on the idea of computing delay from node-to-node transmission to obtain local clock time. A trade-off of accuracy is taken for this protocol to be more efficient.

Reference broadcast synchronization is then discussed. This protocol uses messages from a transmitter to calculate local time. The messages from the transmitter do not include a time stamp; instead, the child nodes in the network record the time when they receive the message and exchange that information with their neighbors [7]. This protocol increases its precision by increasing the amount of messages sent from the transmitter.

Some other protocols like Flooding Time-Synchronization Protocol (FTSP), Time Diffusion Synchronization (TDP), Time Diffusion protocol, and Interval-based synchronization are also introduced in this chapter. As security is an important topic in wireless sensor networks, this chapter will also introduce some of the protocols that attempt to make the time synchronization among the nodes more secure. First to be discussed is the Secure Pair-wise Synchronization [8]. This type of secure synchronization uses a shared key among the nodes that wish to update their clocks. This protocol works with only two nodes in the network. Next, two multi-hop secure synchronization protocols will be discussed. They are Secure Opportunistic Multi-hop Synchronization and Secure Direct Multi-hop Synchronization. These protocols are then scaled to try and secure an entire wireless sensor network against attackers.

Although there are protocols in place to help wireless sensor networks defend themselves against attackers, it will be sufficient to cover some of the attacks brought onto these networks. One such attack is the denial of sleep attack, which is covered in Section 6.5.8. This type of attack denies the node's attempt to go into a low power mode to conserve energy while not communicating with other nodes on the network. This in turns depletes the node's power supply rather quickly and robs its ability to collect and send data back to the gateway for processing.

These topics will be covered in the following sections in this chapter. In Section 6.2, the necessity of time synchronization in sensor networks and its affect on wireless sensor networks is discussed. Section 6.3 describes a basic understanding for time synchronization, which also includes a basic way for synchronization. Section 6.4 introduces several time synchronization protocols. In Section 6.5, some security time synchronization protocols are introduced. In Section 6.6, synchronization protocols in underwater sensor networks are analyzed. Finally, the chapter is concluded in Section 6.7.

6.2 Necessity of Time Synchronization in Sensor Networks

Time synchronization is critical in many sensor network tasks such as object tracking, surveillance, duplicate detection, power-saving duty cycling, or distributed beam-forming. It plays a crucial role also in data integration. All of these applications explain the great deal of attention to the clock synchronization problem in sensor networks, and the large volume of work that has appeared in the last few years. In this section we first analyze the necessity of time synchronization, and then introduce the time synchronization protocols in wired networks. After that we analyze the difference between wired and wireless networks in the design of such protocols, and we also provide the challenges and principals for such protocols.

6.2.1 Necessity of Time Synchronization

Time synchronization is of great importance in sensor networks because physical time plays a key role in many sensor networks applications. Kay Romer and his workmates classify the applications of physical time into three aspects [22]:

6.2.1.1 Applications between Sensor Network and Observer

In many applications, an external observer, such as a human user or a computing system, utilizes sensor networks to conduct tasking, reporting results, and management [22]. The specification of time windows of interest such as "only during the noon" is usually involved in tasking a sensor network. Because a sensor network sends observation results to an observer outside, the observed physical phenomena's temporal properties may be interesting. For example, the number of occurrence of a physical object is important for the observer to relate event reports with the very events [22]. Physical time is also important for determining properties, e.g., speed or acceleration, etc.

6.2.1.2 Applications between Sensor Network and Real World

Data fusion is also a very important process when the nodes of a sensor network observe a physical object [22]. In data fusion, different observations of the same object can be assembled into a coherent estimate of the original phenomenon while these observations are detected by different nodes separately. In such a case, time synchronization is n important for the result. For example, if the proximity of an object is detected by sensors, speed, size, or shape can be obtained by correlating data from multiple sensor nodes [22]. If there are two consecutive nodes N_1 and N_2 that observe a mobile object first and second, then the estimated velocity of this mobile object V_e can be calculated in such function: $V_e = S_d/T_d$, where S_d is spatial distance between N_1 and N_2, and T_d means time distance between N_1 and N_2 (Figure 6.1).

Because lots of instances of a physical phenomenon can happen in a short time, a sensor network's goal is to separate a lot of sensor samples. It is the partitioning of sensor samples into groups that each represents a single physical phenomenon. A key element for separation could be temporal relationships among sensor samples. Correctness and consistency of distributed methods are ensured by temporal relationships among sensor nodes. For example in [24], if the frequency of an observed phenomenon is higher than a sampling rate of sensors, sensor readout needs to

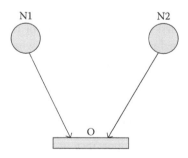

Figure 6.1   Two consecutive nodes N₁ and N₂ see the object 1 in the sequence.

happens at the same time for all sensor nodes to avoid false observations. Moreover, temporal coordination is needed for many actuator nodes.

6.2.1.3 Applications within Sensor Network

In sensor networks, different nodes need to be coordinated. Among different sensor nodes, time is a very useful tool for intra-network coordination. In [25], concurrency control (e.g., atomicity, mutual exclusion), security (e.g., authentication), data consistency (e.g., cache consistency, consistency of replicated data), and communication protocols (e.g., at most-once message delivery) are used in sensor networks. There is a very important instance for concurrency control, i.e., time division multiplexing. Based on assigning time slots to the communicating nodes, multiple accesses to the shared communication media can be reached. This needs the participating sensor nodes to share a common view of physical time.

Some methodologies try to improve energy efficiency by making sensor nodes sleep and wake periodically. Temporal coordination of the sleep periods among sensor nodes may be required in order to ensure the operation of the sensor network.

Another important service for sensor-network applications is temporal message ordering [26]. Time of occurrence such as velocity estimation sketched is used for sorting in many data-fusion algorithms [27]. But, there exist high and variable communication delays so that the ordering of received packets may be different from the ordering of the sent packets. Sensor nodes need temporal coordination when reordering messages [22].

6.2.2 Time Synchronization in Wired Networks

In wired networks, there are mainly two methods for time synchronization, which are Network Time Protocol and Global Positioning System.

6.2.2.1 Network Time Protocol

In Network Time Protocol, there is always an extremely accurate clock in a server. If the client computer/node needs to synchronize with the server, it will send a UDP packet including the required message to request the time information. The server will then return the timing information and thus the computers/nodes can be synchronized when the server returns the timing information [19].

6.2.2.2 Global Positioning System

Global positioning systems (GPSs) are used for wireless devices that are able to communicate with satellites in order to synchronize. Therefore, each wireless device should have a receiver to receive the signals from satellites. The time accuracy of GPS depends on how many satellites the receiver can communicate with at the same time. Line of sight is required for the GPS to communicate with satellite and light of sight is not always available for wireless devices [19].

6.2.3 Comparison of Time Synchronization in Wired Networks and Wireless Sensor Networks

Based on the successful applications of time synchronization methods in wired networks such as Network Time Protocol and GPS, computer scientists expected to transfer these mature methodologies to wireless sensor networks. However, wired networks and wireless sensor networks have so many differences that the techniques or methodologies are not suitable for wireless sensor networks. In wireless sensor networks, nodes are maintained by batteries, which have limited energy and are unable to work as a server to synchronize. Therefore, Network Time Protocol is unsuitable for wireless sensor networks. Meanwhile, GPS is not suitable for wireless sensor networks either. Except for the limitation of energy, the cost of establishing a GPS receiver on each sensor node is too high, especially compared to the chip cost of the node itself. Therefore, the traditional time synchronization protocols for wired networks are impractical to wireless networks due to their constraints in size, power, complexity, etc.

There are several differences in application requirements between wired networks and wireless sensor networks [23,34]. Time synchronization protocols for wireless sensor networks should be designed according to their special requirements:

- *Energy utilization*: There is a major difference between wired networks and wireless sensor networks. In wired networks, there is no limit on energy. However, in wireless sensor networks, all protocols should take into account the limited energy resources contained in sensor nodes
- *Precision*: Stations in wired networks are considered to have better precision. However, in wireless sensor networks, with all kinds of nodes located in different places, it is difficult to achieve this [21].
- *Structure*: In wired networks, there is a time synchronization hierarchy that controls the nodes in the networks in a single, global timescale. Therefore, there is also a system's master clock that determines the time. However, in wireless sensor networks, with so many nodes in different locations, there is no infrastructure. In many applications, sensor networks are deployed in remote, unexploited, or hostile regions. Therefore, they cannot rely on sophisticated hardware infrastructure.
- *Deployment*: In wireless sensor networks, after initial deployment, it is often not possible to access the sensor nodes for hardware or software maintenance. But in wired networks there is no such consideration.
- *Topology*: Stations in wired networks can be configured manually with a regular topology. This is not the case for a sensor network in which network

topologies are dynamic. Due to the deployment of sensors, wireless sensor networks are subject to great dynamics. Meanwhile, sensor nodes can be mobile and they can die because of depleted batteries or environmental influences. New sensor nodes may join anytime and anywhere. All of them can result in the frequent and unpredictable changes in the network topology.

■ *Lifetime*: The synchronized time can be instantaneous or as long as the network lifetime. Synchronization can be global time or local among spatially close nodes, whereas it is difficult to achieve global synchronization in terms of costly energy and bandwidth usage in a large sensor network. A time-based approach is used for aggregating data.

■ *Cost and Size*: GPS is too expensive compared with the sensor nodes. Low cost and small size are design considerations.

■ *Immediacy*: Some applications such as emergency detection (e.g., gas leak detection, intruder detection) need immediacy without tolerating much delay.

6.2.4 Design Challenges for Time Synchronization Protocols in Wireless Sensor Networks

Wireless sensor networks have many applications as they can expand our ability to monitor and interact remotely with physical entities. Sensors can be accessed remotely and placed where it is visionary to deploy data. A synchronization protocol for sensor networks must address the following features of these networks [31]:

■ *Limited Energy*: The efficiency of computing devices is increasing very fast; the energy consumption of the wireless network becomes a chokepoint. Because of the very small size and cost-effectiveness of sensor nodes, thousands of nodes are deployed and this makes it difficult to recharge them. Synchronization must be achieved while preserving their energy to utilize these sensors in an efficient way.

■ *Limited Bandwidth*: Wireless networks consume much less power in data processing than transmitting it. Currently wireless communication is bounded to a data rate of 10–100 Kbits/sec. Bandwidth limitation directly affects message exchange among sensors and it is difficult to achieve synchronization without message exchange.

■ *Limited Hardware*: Sensor nodes are expected to be as small as possible to save energy and costs. Due to their small size, the hardware is very much restricted.

■ *Unstable Network Connections*:
  – The advantage of a wireless network is mobility. The communication of mobile sensors is very limited, and this makes communications among nodes difficult.
  – Topologies change very frequently due to the mobile nature of nodes.

- – Wireless networks suffer from limited bandwidth and periodic connectivity.
- – The wireless medium is unprotected to the external interface and this may lead to a high percentage of message loss.
- ■ *Tight Coupling between Sensors and Physical World*: Wireless sensor networks are designed with tight coupling with the physical world for various applications such as forest fire monitoring, military tracking, and geographical surveillance, and sensors can measure temperatures, light, sound, or humidity. The type of sensors is decided by the applications.

6.3 Basic Conception of Time Synchronization

Nodes in the wireless sensor networks have to maintain a synchronized state to transfer messages efficiently. If synchronization is not maintained, energy consumption could increase due to message retransmission. Also, events could be missed due to nodes being in their low power state at the wrong time. The design goal of wireless sensor networks is to develop a system that has high precision, longer lifetime, larger scalability, and higher throughput.

There are many factors that contribute to synchronization [4]. The most common are

1. *Stability*: This measures how well the local oscillator can maintain its rated frequency. Nodes generally have inexpensive oscillators that have to be accounted for some drift.
2. *Offset*: The time difference between two nodes.
3. *Accuracy*: Time on the local node as compared to the global network time.
4. *Skew*: The frequency difference in the oscillators of two nodes.
5. *Synchronize Frequency*: To achieve this means to regulate the frequency difference between two nodes.
6. *Synchronize Time*: Two nodes have set their time at a given instant to be the same.
7. *Synchronize Clocks*: For nodes to achieve this, they must coordinate their clocks in both frequency and time.

6.3.1 Synchronization of Two Nodes

All nodal synchronization is done by wireless message transmissions. When node A wants to synchronize with node B, it will send a synchronization pulse message at time T1 to node B. This message will contain node A's level and the time T1 when it was sent. Node B will receive the message at time T2. Then, node B sends an acknowledgment to node A at time T3, which contains the level number of node B as well as T1, T2, and T3. By calculating the difference between T1 and

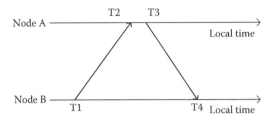

Figure 6.2 Two-way communication between nodes.

T2, node A can correct its clock and successfully synchronize with node B. This is the basic synchronization protocol for wireless sensor networks and is illustrated in Figure 6.2.

The problem with this model is that the transmission time is assumed to be equal from either node. We know that these values are not equal; therefore, they introduce some error in the clock synchronization among the two nodes. There are many factors that combine to constitute the transmission time from node to node. The most common have been divided into four parts by Kopetz [4, 37]:

1. *Send Time*: This is the time it takes the sending node to assemble the message. The sender places the timestamp after the messages so no excess timing offset is encountered.
2. *Propagation Time*: The message needs time to travel from node to node. This is the time that the message has left the sender, but has not reached the receiver. The distance between the nodes has a huge factor in the propagation time.
3. *Receive Time*: This is the time for a node to receive a message from the channel and notify the host of its arrival. This time includes receiving the entire message from the transmission medium.
4. *Access Time*: The node takes time to access the channel. This also includes some time to implement the collision avoidance algorithm by sensing the channel. This is because most wireless sensor networks use the IEEE 802.16.4 ZigBee protocol and all messages are on the same communication channel.

6.3.2 Multiple Node Synchronization

We now extend the two-node synchronization model to multiple nodes. The wireless sensor network can be represented by a spanning tree model (Figure 6.3). We can assume linear runtime for the synchronization algorithm by performing clock updates at the same time for each node on the same level of the spanning tree. The

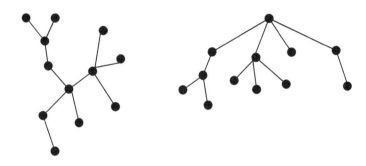

Figure 6.3 **Network layouts: left is physical layout, right is corresponding tree structure.**

root of the tree is the initiating node that starts the synchronization process. It sends a message to all nodes in its range, that is, its child nodes. After those nodes receive the message from the root node, they send the message to their child nodes and the process continues until every node in the tree has synchronized its clock with the root node. Note that wireless sensor networks use omni-directional antennas. In the case of synchronization, when a message is sent from a parent node to a child node, the child node also sends the message back to the parent node. The parent node drops this message because it has no use for it.

There are two types of multiple node synchronization algorithms. The two types are centralized and distributed [1]. Centralized synchronization is just an extension of the two-node synchronization algorithm. The network is constructed into a tree structure, with the reference node acting as the root. The root initializes clock synchronization with its immediate child nodes. Next, the child nodes of the root initialize clock synchronization with each of their child nodes. This process is followed until every leaf node in the network is reached. When every leaf node has finished, the nodes are synchronized. The running time of this algorithm is proportional to the depth of the tree. Every time the centralized synchronization algorithm is run, the tree structure of the sensor network is rebuilt. This allows notification of node movement, newly added nodes, and nodes that have left the network. The resynchronization responsibility lies with the reference node.

Distributed synchronization does not hold the reference node responsible for resynchronization. Instead, any node can induce its own resynchronization. This version of the algorithm does not make use of a spanning tree to organize updates. Alternatively, when a node determines to be resynchronized, it will send a request to the closest reference node. All of the nodes in the path from the reference node to the requesting node will be resynchronized. An advantage of this type of synchronization is that it can keep certain nodes updated more frequently than others.

6.4 Time Synchronization in Sensor Networks

6.4.1 Timestamp Synchronization

Timestamps are used in some distributed wireless sensor networks to synchronize the clock on each of the nodes [10]. Each of the messages transferred in the network has a timestamp with a local system clock. Timestamp is more likely to be a boundary of time when a certain event occurs instead of an absolute time. Whenever a node sends a new message, the latest timestamp is processed according to the local clock and attached to the message. Meanwhile, drift rates between nodes in the network are considered for better approximation of the boundary of timestamp.

The drift between local and network clocks is not directly known, but an estimated value. This is performed by taking the time of the local clock and the network clock immediately at two moments [10], i.e., first the local clock, then the network clock, and finally the local clock so that the bound is obtained (the first time taken is a lower bound and the last time taken plus the clock granularity is the upper bound) [10].

Some of the advantages of online time stamping are that no local clock is changed. This allows instant accurate time stamping even if the network configuration changes [10]. Another advantage is that online time stamping gives more accurate time estimations than protocols that use fixed drifts to estimate clock offsets in the nodes. Also, this protocol can be performed with few calculations. It uses no tables to store time equivalences [10].

6.4.2 Time-Sync Protocol for Sensor Networks

The goals of a wireless sensor network's time synchronization protocols are as follows: to be accurate, lightweight, flexible, and comprehensive. Time-sync (Tsync) [2] protocol is an example of such type of protocols.

One of the drawbacks in communication over a wireless media is the likelihood of packet collisions. In a single-channel wireless network, each node (something similar to the CSMA/CA protocol) first sends a request packet for sending data, and then waits for an acknowledgment before sending data. This protocol minimizes but does not eliminate packet collisions. The sender will sense the channel for some time before trying to send a message. Tsync uses multiple frequencies in the available channel to reduce the possibility of collisions, which is different from some popular wireless protocols such as IEEE 802.11.

To make TSync flexible, designers made it bidirectional [2]. It has two variations of clock synchronization: push and pull. The push component of TSync enables "reference nodes" to initiate the resynchronization of the wireless sensor network. Multiple reference nodes can exist in the sensor network. In case of multiple reference nodes, a shortest path tree is formed around each reference node. If a child node has several reference nodes that are the same distance from it, that node will

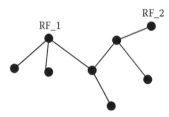

Figure 6.4   Multiple reference nodes.

resynchronize from the reference node that contacts it first. An example of a network with multiple reference nodes depicted in tree form is shown in Figure 6.4.

The other variation of TSync is called pull-based [2]. This version allows independent nodes to initialize their own resynchronization with a reference node. When a node wants to resynchronize, it sends a request to its immediate neighbors. The neighbors propagate the request until it reaches a reference node. The first reference node that is located will be selected for synchronization. If, for instance, the topology of the sensor network is known, the request could be targeted to a specific node. In another instance, if the path to the reference node is unknown, and the original request returns no result, it can simply use its neighbor to resynchronize.

6.4.3 Global Synchronization

Most wireless sensor networks require that their nodes agree on a common global time. Having a global clock will help nodes in the network be more efficient in transmission of data and lets them conserve energy by knowing when to go into a low power state. The drawback for using a global clock to "wake up" nodes is that more complex communications have to be in place to keep the node clocks synchronized within a bounded limit. In this section, we will discuss three ways to implement global clock synchronization in the wireless sensor network. The three ways are the all-node-based method, the cluster-based method, and the fully localized diffusion-based method [3].

6.4.3.1 All-Node-Based Synchronization

In this approach, we take for granted that the clock cycle on each node is the same. We also assume that the clock tick cycle is longer than the transmission time. This is an important assumption because a lower clock frequency consumes less energy than a higher clock frequency [3].

The initial setup finds a cycle that passes through each node that needs to be synchronized in the network at least once (Figure 6.5). The initiating node then sends a message to each node in the cycle. This message has the first node's current

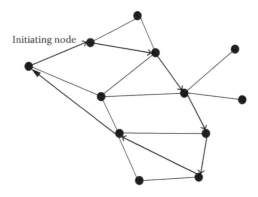

Initiating node

Figure 6.5   All-node-based synchronization cycle.

time. When the other nodes in the cycle receive the message, they record their local time and their order in the cycle. When the initial node receives the message back, it sends out another message with the start time (t_s) and ending time (t_e) of the previous message cycle. For each node to adjust its local clock (t) to the global clock, they use the equation $t = t - t_i + t_s + m$ where t_i is the node's local time. After all of nodes in the network receive the second message and compute their new time, the network is globally synchronized.

6.4.3.2 Cluster-Based Synchronization

The problem with the all-node-based synchronization algorithm is that every node in the network has to participate in the same synchronization session. The cluster-based synchronization protocol attempts to fix this problem.

The cluster-based algorithm works the same way as the all-node-based synchronization, but only with the head cluster nodes. Head nodes synchronize themselves first, and then the child nodes in each cluster synchronize with the head node. This method of synchronization allows some clusters to synchronize independent of other clusters. This also helps conserve energy in that some clusters may have to synchronize more often than others, leaving the less frequently updated nodes in an energy saving state.

6.4.3.3 Fully Localized Diffusion-Based Synchronization

The synchronization methods described in Sections 6.4.3.1 and 6.4.3.2 do not work well in a very large-scale wireless network. They have single points of failure, in which case some nodes will never be synchronized again. The local diffusion-based protocol is an attempt to fix this problem.

This method works like the time synchronization between two nodes explained in Section 6.3.1. Nodes exchange local times with neighboring nodes. This diffusion occurs until all sensor nodes in the network have updated their local clock. One of the benefits of this time synchronization protocol is that there is no point of failure. If a node wishes to update its local clock, it can use any of its neighbors. No single node initiates a global synchronization of all nodes in the network.

There are two versions of the localized diffusion-based algorithm: the synchronous diffusion algorithm, and the asynchronous diffusion algorithm [3]. The synchronous diffusion algorithm has to perform the operations to obtain a bounded global clock in a set order. All of the nodes that are finished synchronizing with their neighbor must wait for the other nodes that have not finished. The asynchronous diffusion method does not have the time constraint of letting all other nodes in the network finish their synchronization before starting another round. This method allows any node in the network to initiate its own time synchronization any time it needs.

6.4.4 Adaptive Clock Synchronization

To implement causality, wireless sensor networks must synchronize their time with the fastest clock [4]. No node will be permitted to decrease their clock time for synchronization. In a given situation where one event happens before another event, the first event must happen at an earlier time than the second event. If nodes of the wireless network are allowed to adjust their clocks backwards to synchronize with a global clock, the previous statement could be violated. For instance, we assume that node one has a local time of 10 and node two has a local time of 12. Node two sends a message (event one) and then synchronizes its clock with node one by reducing its local clock time. After that, node two sends another message (event two). The clock offset of the two nodes could place a timestamp on the second message, which seems that it was sent before the first message.

6.4.4.1 Related Synchronization Protocols

There are many implementations of the clock synchronization protocol [4]. A simple one-way message is enough in the case where the latency is lower than the accuracy that each node needs. The client will measure the round trip time to calculate one-way latency [6]. Listed below are a few examples that are used in specific networks; however, they may not be sufficient in wireless sensor networks.

■ *Wireless Clock Synchronization Protocols*: These protocols are exclusively applied in wireless or ad-hoc networks. Instead of adjusting the clock on each node in the network, this protocol adjusts the timestamp. From a source to a destination, a message is passed along each node and has its timestamp changed to match the node's local clock. Error is introduced in the timestamp along the path of nodes, with more error introduced as the path gets longer. The 802.11 wireless protocol implements this kind of clock synchronization, and it has beed found to not be scalable to a large number of nodes due to the increased error introduced by each hop.

■ *Receiver–Receiver Synchronization*: This type of clock synchronization is done among individual nodes in the network. Any node that needs a clock update can make an initiation.

■ *Probabilistic Clock Synchronization*: A type of clock synchronization that tries to correct unacceptable update attempts. One attempt is to interpret the clock of the master node repeatedly until a reply is accepted. This algorithm takes $2/(1 - p)$ messages from the master node, where p is the probability of losing the message. This algorithm is repeated n times to reach a probability of synchronization equal to $1 - p^k$. This protocol wastes excess amounts of energy by sending large amounts of messages to synchronize the nodes.

6.4.5 Delay Measurement Time Synchronization

Delay measurement time synchronization (DMTS) is a protocol that trades accuracy with efficiency [5]. It is based on evaluating different forms of delay while updating local clocks in the wireless sensor network. The Berkley motes platform and TinyOS was used in implementing this algorithm [5]. Clock synchronization is dependent on the types of oscillators on the motes. DMTS intends to use small computation complexity and low memory occupation to conserve energy in the nodes. Also, this protocol aims to be flexible to different network topologies.

To initiate the resynchronization, a leader node is selected as the time master and broadcasts its time. When the message reaches the nodes in the range of the leader, they set their clock to match the leader's clock plus the time delay it took to receive the message. After all of the nodes receive the message from the leader; their time will be resynchronized bounded by the efficiency of the delay measurements along the path. Delay is composed of the factors affecting transmission time from node to node described in Section 6.3.1.

The precision of delay measurements along the path limits the synchronization accuracy. But it is energy efficient since only one message is required.

It is not good for multiple hops. DMTS uses the concept of a time-source level to identify the network distance of a node from the master. First, a time master node is chosen. This node is labeled at level 0. All of the immediate child nodes of the time master are labeled at level 1. Other nodes in the network at level n are labeled level $n + 1$. In this way, each node has a level number and knows how many hops away that they are from the time master node. The time master node will occasionally send a message with its local time. Each node that receives a message from the time master node will update its time and broadcast the time message again. These nodes broadcast the time message only once. When the message propagates through the network other nodes will do resynchronization according to the one of their parent nodes that is the closest to master node. It then sends a message with the time to its neighbors. This happens until all of the nodes in the network have synchronized their time with the time master node. Also, this protocol produces a minimum amount of traffic by having each node send a time message only once. The traffic load is directionally proportional to the number of nodes in the network [5].

The time master node can be any node in the network. Any selection algorithm can be applied to choose the master, but in most cases, simple voting algorithms are used. This algorithm is run at any time when a mater node is not present in the network [5]. Best practice is to have a base station selected as the time master. The reason for this is that the base stations are typically not mobile and are not maintained by batteries. A base station also typically has more processing power than the sensor nodes in the network. Choosing a base station to be the master time node is preferred, but not mandatory. Any node in the wireless sensor network can be elected to be the time master node.

6.4.6 *Reference Broadcast Synchronization*

Reference Broadcast Synchronization (RBS) protocol [28,31] is so named because it exploits the broadcast property (transmitting property) of the wireless communication medium [32]. Instead of sending a message with a timestamp, the nodes use the message's time of arrival to compare their local clocks [6]. One of the important constraints of using this protocol is that it requires a physical broadcast channel. It will not work with a wireless sensor network that employs direct point-to-point links [6].

In RBS, two receivers located within the transmission range of the sender receive the same packet at approximately the same time. If each receiver records the received local time, all of receivers can synchronize with each other with a high degree of precision [31]. RBS [31] protocol utilizes several messages from a given sender to calculate offset and skew of the local clocks relative to each other using the concept of time-critical path (a path of a message that contributes to non-deterministic errors in a protocol).

RBS [31] protocol is relatively accurate if it is for single hop without errors. Some of the advantages of RBS are [6,31]:

1. The major sources of latency can be removed from the critical path. This is done by using the broadcast channel instead of point-to-point connections to synchronize nodes with one another.
2. This protocol uses multiple broadcasts to allow a more precise synchronization among the nodes. This allows the nodes to be synchronized less often and lets them go into a sleep state when not being used.
3. Nodes in the network can get a local time reference, even if they are not synchronized with the entire wireless network.
4. Clock offset and skew are estimated independently of each other.
5. The sources of error (send time and access time) are removed from the critical path by decoupling the sender from the receivers.
6. Multi-hop support is provided by using nodes belonging to multiple neighborhoods (i.e., broadcasting domains) as gateways.

6.4.7 Flooding Time-Synchronization Protocol

Flooding Time-Synchronization Protocol (FTSP) is a sender-to-receiver synchronization algorithm. A root node is selected and all other nodes are synchronized to the root node. The root node transmits the time synchronization information with a single radio message to all its children nodes. The content of the message is the root node's timestamp of the global time at transmission. The receiver node notes its local time when it receives the message. The receiver can estimate the clock offset by having both the sender's transmission time and its reception time. The message is MAC layer time stamped, on both the sending and receiving sides. FTSP uses linear regression for the compensation of clock drift, which improves the precision.

FTSP is designed for large multi-hop networks. The root is selected dynamically and reelected periodically. It is responsible for keeping the global time of the network. The receivers synchronize themselves to the root node and organize in an ad-hoc way to communicate with all of other nodes about their timing information.

The main advantages of FTSP are flexibility for dynamic topology changes, robustness for node and link failure, and MAC layer time stamping for precision. It uses the flooding of synchronization messages to combat link and node failure. The flooding also supports the ability for dynamic topology changes. The protocol enables the root node be periodically reelected, so that it is robust for node and link failure. FTSP also provides MAC layer time stamping, which greatly increases the precision and reduces jitter. This will eliminate all but the propagation time error. It utilizes the multiple time stampings and linear regression to estimate clock drift and offset.

Figure 6.6 gives the establishment process of the transmission of data packets with FTSP. There is a preamble, then sync bytes followed by the data, then finally the CRC (Cyclic Redundancy Check). The solid lines indicate the bytes in the buffer while the dashed lines in the figure indicate the actual bytes in the packet. When the sender is transmitting the preamble bytes, the receiver adjusts to the carrier frequency. Once the sync bits are received, the receiver can calculate the bit offset needed to accurately recreate the message. The timestamps are located at the boundaries of the sync bytes [19].

6.4.8 Time-Diffusion Synchronization

Sensor nodes in Time-Diffusion Synchronization Protocol (TDP) can obtain a local time that is within a small bounded time deviation from the network-wide "equilibrium" time [31]. The protocol needs to be applied periodically to reduce clock skews in alternating active and inactive phases. In active phase, there are multiple cycles, each cycle lasting a duration τ [31]. In each cycle, a subset of the nodes is elected as the masters by an Election/Reelection Procedure (ERP) [31].

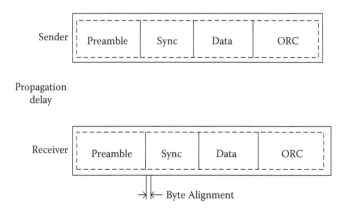

Figure 6.6 The transmission of data packets in FTSP.

Each master initiates a diffusion of timing messages; these messages effectively make a tree-like multiplicative structure dynamically for each diffusion in the network. Nodes without leaf in this tree are the nodes that propagate the timing messages, and are termed as "diffused leaders" [31]. Then these diffused-leader nodes are elected by ERP. Thus, it may happen that a node does not qualify to be a diffused leader node, and therefore it will not propagate the diffusion. The goals of the ERP are [31]:

■ To eradicate outlier nodes whose clock variation is higher than the threshold function based on a specific type of variance calculation, known as the Allen variance. This variation is determined by exchanging messages and calculating deviations among the pairs of adjacent nodes using a Peer Evaluation Procedure (PEP) [31].

■ To accomplish load distribution among the nodes as the roles of masters, diffused leaders put a greater demand on the energy resource. The load distribution is attained by taking turns at being the master, based on factors such as the available energy level being above a tunable threshold. In each cycle, the diffusion of timing messages helps converge the local times, and reach a common notion of the system-wide time.

This synchronization protocol has several advantages [31]:

■ This protocol is geared towards mobility.
■ This protocol achieves an "equilibrium" time across all nodes, which is computed by an iterative weighted averaging scheme. This involves all the nodes in the synchronization process.

- The diffusion does not trust on a static level-by-level transmission. This non-dependence on a static structure provides flexibility and fault tolerance.
- The protocol is tolerant of message losses.
- The TDP protocol provides synchronization even without external time servers.
- It has hierarchical structure but still it can be neutralized by having multiple master nodes, which are distributed across the network.

6.4.9 Interval-Based Synchronization

The objective of interval-based synchronization is to sustain the guaranteed upper and lower bounds on time offsets and keep the distance between the two bounds as small as possible [8]. The bounds on time are given by T^l and T^u. These bounds are the limits on the variation compared to the real time. The equation $\Delta T = T^u - T^l$ is the uncertainty [8]. Nodes trade their current time bounds with each other when they make communication. Of these bounds the two nodes pick the best ones.

Another type of interval-based synchronization is called the Back-Path Interval-Synchronization Algorithm (BP-ISA) [8]. This algorithm works like the previous version except that every node stores the optimized bounds from every other node that it has previously communicated with instead of only storing one pair of optimized bounds. This version improves on the average case of the original version. The overhead for BP-ISA is directly proportional to the number of nodes in the wireless network.

6.4.9.1 Communication Patterns

Nodes in the sensor network need to know when to communicate. A communication pattern propagates this information throughout the network [8]. There are two types of network layouts for communication patterns.

- The network can resemble a tree structure. In this case, the root sends a synchronization message to all of its children, which in turn synchronize with their children. This happens until all of the nodes in the wireless sensor network have synchronized with the root.
- Nodes in the network randomly synchronize with their neighbors. There are no root nodes as any node can initiate synchronization whenever it wants. This is referred to as a random communications topology.

Both topologies have advantages and disadvantages. The tree topology has the route path to any node at any time but has to accommodate for a large overhead of messaging if the sensor nodes are mobile. The tree has to be rebuilt repeatedly with a mobile ad-hoc sensor network. The random communication topology does not deal with these situations. This topology works like the distance vector protocol in infrastructure networks.

Energy consumption is directly related to the quantity of messages sent by the nodes in the network. In the tree-based topology there are two layouts: breadth-first (star) and depth-first (classic tree). Message passing in the star topology is evenly distributed if messages are broadcasted. If messages are transmitted point to point, the root node has significantly more transmissions than the other nodes in the network [8]. In the random communication layout, message transmission is uniform because of the arbitrary selection of communication between nodes [8].

6.5 Secure Time Synchronization

Nodes in a wireless sensor network depend on the most accurate time synchronization they can obtain to run their respective protocols. The effect would be harmful if an intruder could interfere with the messages among the nodes. This could cause devastation in the network with lost packets, jammed signals, or false messages.

There are many time synchronization protocols for wireless sensor networks, but most of them do not implement security measures. Most wireless sensor protocols are designed for a non-hostile environment, so security measures are not used. However, there are many applications requiring secure measures to communicate and synchronize. The following are some of the types of attacks that can be performed on a wireless sensor network [7]:

■ The attacker can modify messages. This type of attack changes the content of the message so that the receiver will achieve the wrong information.
■ Another type of attack is called the pulse-delay attack. This attack jams the message channel from the sender to the receiver as shown in Figure 6.7. The attacker in turn receives the message from the sender and after some time, sends it to the receiver. This type of attack affects the propagation delay between nodes and causes incorrect clock computation.
■ The attacker can modify ACK messages from the receiver back to the sender.

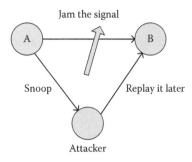

Figure 6.7   Description of the pulse-delay attack.

Even though the following protocols are not usually implemented, it will be sufficient to give a brief overview of them here.

6.5.1 Secure Pair-Wise Synchronization Protocol

Nodes using this protocol share a secret key. This prevents an attacker from assuming the identity of either of the nodes. This type of security does not prevent pulse-delay attacks. To fix this problem the nodes have a calculated maximum message delay. If an attacker is using the pulse-delay attack to skew the time synchronization, it will be caught by comparing the delay of the received message to the maximum expected message delay [7]. If the delay is more than expected, the synchronization calculation between those two nodes is terminated.

It should be stated that there could be an attack that is not caught by the nodes. This attack is executed by altering the time difference between the nodes without getting detected. Using the Secure Pair-Wise Synchronization protocol, the maximum time difference an attacker can achieve without being detected is around 20 μs.

When there are no attack attempts the Secure Pair-Wise Synchronization protocol performs as efficiently as other non-secure wireless senor network protocols.

6.5.2 Secure Opportunistic Multi-Hop Synchronization

This is an extension of the previous protocol. The difference is this protocol is for multiple hops. The two nodes still share a private key, but the intermediate nodes between them do not share this key. This protocol is also resilient against the pulse-delay attack by calculating the maximum delay time from each node through the multi-hop path that is connecting them. This delay time is calculated by summing all of the delays between any two nodes in the path. The maximum expected delay is higher than the sum of these pair-wise delays to account for increased traffic or a specific part of the wireless network. A wireless network implementing this protocol needs $2n$ packet transmissions for nodes that are n hops apart.

Again, it should be noted that attackers can achieve disorientation without getting noticed. In this Secure Opportunistic Multi-hop Synchronization, attackers can skew clock updates by up to 50 μs [7].

6.5.3 Secure Direct Multi-Hop Synchronization

This multi-hop protocol requires only the neighboring nodes to share a private key, not the two end nodes. Using this protocol, there is a larger overhead than the earlier protocols. Every packet has to contain the state information of every node that it passes through from source to destination. Also, being the same as the protocol in Section 6.5.2, this protocol takes $2n$ packet transmissions for nodes that are n hops apart.

Attacks on this protocol can skew the synchronization of two nodes up to 50 μs [7].

6.5.4 Secure Transitive Multi-Hop Synchronization

This protocol is also an extension of the protocol in Section 6.5.1. This protocol differs from the others in that it does not need to compute a maximum delay time from a sender to a receiver.

The way that this protocol has validation locally on every link instead of over the entire path gives attackers a chance to introduce errors between any two nodes [7]. This protocol requires $3n$ packet transmissions for nodes that are n hops. A wireless network using this protocol can have an attacker go undetected while introducing errors between any two nodes. The total amount of clock skew that an attacker can introduce in this protocol is 120 μs.

6.5.5 Group Synchronization

Sensor networks require nodes to have their clocks synchronized. In this section two group synchronization protocols will be examined. One is the Lightweight Secure Group Synchronization [7]. The other is Secure Group Synchronization. Both of these protocols can detect packet modification and pulse-delay attacks from outside attackers. Secure Group Synchronization can also detect fake timing information by internal attackers, but costs are higher with more message transmissions.

The first protocol discussed will be the Lightweight Secure Group Synchronization. This protocol does not require a leader election algorithm. The first node to sense an event can generate the messages to initiate time synchronization. First, a node initiates a synchronization request to which the other nodes reply with their ID and challenge nonces. Then the initiating node replies with a broadcast message to all of the other nodes that includes their MAC addresses and ID. The MAC addresses are used to authenticate nodes that are synchronizing [7]. This protocol uses $n + 1$ messages to synchronize a wireless sensor network with n nodes. Also, this protocol can resist message modification and pulse-delay attacks [7]. It is not, however, resilient to attacks from compromised or malicious internal nodes.

The other protocol, Secure Group Synchronization, is resilient to external and internal attacks. This protocol first does a simple consistency check. This is a triangular consistency check that works by traversing a set of nodes that have performed pair-wise synchronization with each other [7]. The edges of the triangle are the offset between each of the two nodes. The cycle computes the sum of all of the edges in the group and compares this number to a threshold that is close to zero. The ideal sum of the offsets would be zero, but this may not be the case because of clock drift or skew error [7]. The threshold is in place to consider these types of errors. In this protocol every node broadcasts its ID and nonces. Then every member broadcasts another message to reply their response to nonces of the other nodes. It also contains the MAC address of the receiving nodes used to authenticate. Next, each node computes threshold verifications on all of the computed delays from it to surrounding nodes. This provides pulse-delay resistance from external attackers [7]. In this protocol, the number of messages needed to synchronize a wireless sensor network with n nodes is $3n$. It uses multiple pair-wise synchronizations at the same time. No leader node is needed to initiate the Secure Group Synchronization protocol. Synchronization is done with the node that has the fastest clock in the network. This node is made known after the protocol is run [7]. This makes the protocol resistant to a single point failure. Also, the triangular consistency check will identify any internal attacker by failing the check and aborting the synchronization process [7].

6.5.6 Secure Network Synchronization

A wireless sensor network using a secure time synchronization protocol can achieve global network synchronization in two stages [7]. First, like most ad-hoc wireless protocols, a tree topology is laid out with a reference node as the root. Then each node down synchronizes with the root. This pair-wise synchronization continues until each node in the network has synchronized itself with the root node. The nodes in the network use the Secure Pair-Wise Synchronization protocol to update their clocks. This provides a resistance to external attackers. The problem with this is how to synchronize the entire wireless network while having some compromised internal nodes. One possible solution to this problem is to use disjoint multiple paths to synchronize nodes to the root node [7].

6.5.7 Attack Resilient Synchronization

There are many attacks that can cause sensor networks to become useless in providing quality data. One of the types of attacks is called a masquerade attack [9]. The idea behind such an attack assumes that there are three nodes that want to communicate to synchronize their clocks. Node A will send a broadcast message to nodes B and C with its timestamp to initiate the synchronization. Unknown to node B, node E has intercepted the broadcast from node A and changed the message. Node E then sends the message to node B with the adjusted payload. In this case, node B has been compromised of an accurate clock update [9]. The second type of attack is called the replay attack [9]. This type of attack works like the one explained earlier, but also repeats old packets to synchronize other nodes to an incorrect time. The third attack is called the message manipulation attack [9]. In this type of attack, the malicious node will do anything it can to disrupt the synchronization process between nodes. Shown in Figure 6.8, the malicious nodes can drop packets, modify packets, and even create new packets to interrupt communication between nodes in the wireless sensor network.

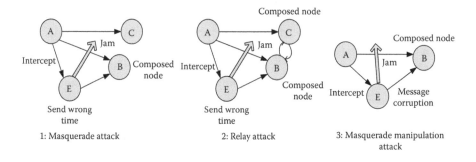

Figure 6.8 Three types of attack.

Adding encryption may help with these types of attacks by authenticating messages from sending nodes. Sequence numbers or a nodal ID can be added for validation of messages [9]. The purpose of the attack-resilient protocols is identifying the malicious nodes and how to accomplish a secure synchronization within the wireless sensor network.

6.5.7.1 Generalized Extreme Studentized Deviate Protocol (GESD)

This protocol is used to identify outliers, which are observations that stray from the normal operations so much that they are called to attention to be malicious [9]. Once GESD finds potential outliers it excludes them to try and acquire a more accurate assessment of the time offset to produce a better time synchronization.

GESD uses two parameters to find multiple outliers. The parameters are r and λ_i, where r is the anticipated number of outliers and λ_i is the two-sided 100 * a percent critical number [9]. Choosing r is important in GESD. If the number picked is smaller than the actual number of outliers in the network, not all of them can be found. Otherwise, if r is picked too high, energy is wasted [9]. The GESD protocol sets the r value less than or equal to the number of synchronization paths in the network. GESD uses this information to find candidate outliers. These candidates are then compared with given values of expected time offsets to determine whether they are outliers or not [9]. The goal of this protocol is to synchronize the sensor network in the presence of attackers and malicious nodes.

6.5.8 Denial of Sleep Attack

The wireless sensor network denial of sleep attack is a division of the denial of service attacks associated with wired networks [11]. These denials of sleep attacks prevent the node's ability to go into a standby state to reduce power consumption while not transmitting data or re-synchronizing its clock to the network time. The denial of sleep attacks can be categorized into three sets [11]:

■ *Service request power attack*—This type of attack continually uses a valid request to intentionally deplete the node's energy source.
■ *Benign service attack*—Sends a request to the node that initiates a power demanding process that drains the energy supply.
■ *Malignant power attack*—Propagates to the device and changes existing programs to use more energy than they would otherwise.

The transceiver on the wireless sensor is the device that consumes the most energy. This device has to be the most regulated to consume energy in the node.

One of the attacks is done on the link layer. This attack is called a link layer collision [11]. When nodes need to synchronize their clocks, the only way they can do this is to send messages to their neighbors. Messages propagate from a root node to the node that needs to synchronize, and the node then gets a new local time. This message transfer has to be done using the transceiver because the node has no other way of communication. If nodes are to protect themselves against denial of sleep attacks, they should implement security measures at the network or MAC layer. Protecting these levels of communication will be best in monitoring unwanted usage of the radio [11].

6.5.8.1 Sensor Security Components

Nodes in a wireless sensor network lack the storage capabilities to retain encryption keys for every link. The processing and storage abilities are limited on each node. There are tools especially for preserving security and resisting attacks from outliers [11].

One of the tools especially made for the link layer is TinySec [11]. This software application provides components for security, authentication, and encryption. TinySec works with TinyOS to provide link layer security without significantly increasing energy consumption. It replaces the CRC field in the link layer frame with message authentication code. TinySec has the capability to operate in three different modes: open, authentication mode, and authentication-encryption mode [11].

A second tool that is similar to TinySec is ZigBee, which is the IEEE 802.15.4 protocol. Motes that use this protocol have the ability to execute 128-bit Advanced Encryption Standard (AES) using hardware instead of software [11]. One hundred twenty-eight-bit AES has the ability to produce 3.4×10^{38} different keys. It is estimated that if a DES key generator were able to discover one key per second, it would take 149 trillion years to crack a single 128-bit AES key using the brute force method. Implementing this directly on the hardware instead of in software frees the limited memory and CPU to handle other processes.

Security Protocols for Sensor Networks (SPINS) was also developed to secure message transition and node authentication [11]. This application uses only 8 bytes of message overhead to implement data confidentiality and two-party data authentication, and to authenticate data freshness. Although SPINS addresses many security issues, it does not address the denial of service attacks like the denial of sleep attack [11].

These security measures do not help much in the event of a denial of sleep attack. This type of attack forces a node to stay active and process packets, even though the packets they receive contain useless or no data [11]. Most nodes have three types of operation, which are encrypting the entire message, encrypting only the data in the message, and no encryption at all on message transfer. Nodes must be active to receive encrypted messages. They stay awake to receive the entire message before decrypting it and determining if it is a relevant message. If the data is the only part of the message that is encrypted the attacker can produce a legitimate header to force the node to process the message [11]. If the message is unencrypted, the node will still process the message before discarding it.

6.5.8.2 MAC Layer Secure Protocols

S-MAC (Sensor MAC) [11] divides a time frame into sleeping and listening periods. The listen period is used to synchronize its clocks with other nodes and for message transfer. The sleep cycle is fixed before the sensor network is deployed. This cycle cannot be changed and therefore protects against the denial of sleep attack, which, however, makes the network inflexible to increased traffic [11]. This protocol can keep the node available for an average of 63 days while dealing with a denial of sleep attack.

Timeout MAC (T-MAC) [11] is built from the S-MAC standards. The difference with this protocol is that the sleep period can dynamically change due to network traffic. The nodes sense the wireless channel for activity and wait on a timeout. This protocol is susceptible to broadcast attacks [11]. If broadcasted messages are repeatedly received, the nodes will deplete their energy source more rapidly than S-MAC. This protocol can sustain the node's availability for an average of 108 days while dealing with a constant denial of sleep attack. This is less than the perfect time that S-MAC sustains whether or not a denial of sleep attack is present. The T-MAC protocol gives an average of 295 days of node availability with no attacks and no network traffic [11].

Berkeley MAC (B-MAC) [11] allows nodes to adjust their own sleep schedule. The nodes use a low power sensing of the channel when they come out of standby. If the node senses activity, it will become active to receive and send packets. B-MAC, like T-MAC, is susceptible to broadcast attacks [11]. The nodes implementing the B-MAC protocol can stay useful for up to 87 days in a persistent denial of sleep attack, and for 244 days when the network is completely idle.

The last MAC protocol is the proposed Gateway MAC (G-MAC) protocol [11]. In this protocol, nodes elect a gateway node to equally distribute energy requirements among all nodes. This happens every so often to accommodate for failing gateway nodes and to distribute the duty of the gateway to others. Nodes in the network only respond to the gateway node, which means all messages have to be authenticated before being sent. This protocol can sustain nodal lifetime for an average of 478 days in the presence of a constant denial of sleep attack. This is compared to a nodal lifetime of 480 days when the network is idle and there is no denial of sleep attacks.

Nodes in wireless sensor networks intend to stay idle for most of the time. They need to communicate only when an event is detected or when they need to synchronize with their neighbors. Denial of sleep attacks disable the nodes to go into a low power state and therefore shorten the lifespan of the node. These attacks are mostly implemented on the link layer, which directly attaches to the transceiver. The transceiver is the most power consuming part of the hardware in the wireless sensor.

6.6 Time Synchronization in Underwater Sensor Networks

In [35,36], time synchronization in Underwater Acoustic Networks (UANs) has been considered, and UANs differ from terrestrial sensor networks since they have long and variable propagation delay and mobility so that time synchronization becomes difficult to achieve. Propagation speed of underwater acoustic signal is only about 1500 meters per second, which is much lower in several orders of magnitude than the radio propagation speed, i.e., light speed, 3×10^8 meters per second.

In [35], a three-dimensional, scalable UAN time synchronization scheme was proposed to achieve both horizontal and vertical clock synchronization to overcome the long propagation delay, where a horizontal level means the same water depth and a vertical level means an approach from the bottom up to the surface. The paper in [35] also considers security issues about the time synchronization correlation test and statistical reputation trust model to detect outlier timestamps.

In [36], the authors consider effects of node movements on underwater time synchronization. Since an underwater node can move out of and into another node's range frequently [36], no time synchronization is necessary if the timestamps of the received data packets are within the tolerance so that the underwater network does not need to perform global time synchronizations periodically, which reduces the time used to synchronize clocks among sensor nodes.

6.7 Conclusions

This chapter has demonstrated how the local clocks of individual nodes in a sensor network can be synchronized. It reviewed different ways in which nodes communicate, either with themselves or back to a master node to update their clocks. Also, many time synchronization algorithms have been reviewed.

The problems leading to nodal synchronization were covered. Some of those problems were local oscillator drift, clock skew, errors in synchronization messages due to delays, and computational complexity in synchronization of multiple hop wireless sensor networks.

Global synchronization was discussed as usually an enhancement of the pairwise synchronization protocols. These protocols attempt to synchronize the entire network with a common clock. The clocks on the individual nodes in the network are allowed to be off by a set of upper and lower bounds.

Security for wireless sensor networks was covered. These security measures were implemented to defend against malicious attacks and compromised internal nodes. These security measures were not as effective in dealing with denial of sleep attacks. These types of security protocols were mainly targeted towards nodal authentication, message integrity, and message encryption.

The denial of sleep attack does not attempt to recover encrypted messages or authenticate nodes. Its sole purpose is to deprive nodes in the wireless sensor network of going into a low power state. There are security protocols specifically made for these types of attacks.

An enhancement to the problem of global synchronization would be to use cellular towers to transmit beacons with a timestamp attached. If the wireless sensor network was located in the cell, the tower could send messages to the root node or all of the nodes to keep their clocks updated. This scheme for synchronization would work like the synchronization method using GPS satellites. The advantage is that the nodes would not have to be in a direct line of sight to the tower. Nodes can be indoors and still receive beacons from the cellular tower. This would also reduce the amount of energy needed in the network because of less message transmission between nodes to keep their clocks synchronized. The drawback would be that the nodes, if they all communicated with the cellular tower, would need different circuits for receiving messages. They would need hardware for local message transmissions and hardware for receiving messages from the base station. Both are needed because each transmits in different protocols and frequency bands.

Synchronization is an important factor in wireless sensor networks. This field will be studied more strenuously as future devices will depend more heavily on wireless communication.

Acknowledgments

This work is supported in part by the U.S. National Science Foundation (NSF) under the grant numbers CCF-0829827, CNS-0716211, and CNS-0737325.

References

1. J.V. Greunen and J. Rabaey, "Lightweight Time Synchronization for Sensor Networks," Proceedings of the 2nd ACM International Conference on Wireless Sensor Networks and Applications, Sept. 2003, San Diego, CA.
2. H. Dai and R. Han, "TSync: A Lightweight Bidirectional Time Synchronization Service for Wireless Sensor Networks," ACM SIGMOBILE Mobile Computing and Communications Review, 2004, 8(1):125–139.
3. Q. Li and D. Rus, "Global Clock Synchronization in Sensor Networks," Proc. IEEE Conf. Computer Communications (INFOCOM 2004), Vol. 1, pp. 564–574.
4. A. K. Saha, D. B. Johnson, and S. PalChaudhuri, "Adaptive Clock Synchronization in Sensor Networks," 3rd International Symposium on Information Processing in Sensor Networks (IPSN '04), April 2004.
5. S. Ping, "Delay Measurement Time Synchronization for Wireless Sensor Networks," Intel Research, IRB-TR-03-013, June 2003.
6. J. Elson, D. Estrin, and L. Girod, "Fine-Grained Time Synchronization Using Reference Broadcasts," in Proceedings of the Fifty Symposium on Operating Systems Design and Implementation. University of California, Los Angeles. 2002.
7. S. Capkun, S. Ganeriwal, C. Han, and M. B. Srivastava, "Secure Time Synchronization Service for Sensor Networks," University of California, Los Angeles. ACM. 2005.
8. P. Blum, L. Meier, and L. Thiele, "Interval-Based Clock Synchronization Is Resilient to Mobility," Second IEEE International Conference on Mobile Ad Hoc and Sensor Systems (MASS 2005), Washington, 7–10 Nov. 05.
9. G. Cao, H. Song, and S. Zhu, "Attack-resilient Time Synchronization for Wireless Sensor Networks," Ad Hoc Networks 5(1): 112–125 (2007) [DBLP:journals/adhoc/SongZC07].
10. O. Bezet and V. Cherfaoui, "On-Line Timestamping Synchronization in Distributed Sensor Architectures," IEEE Real-Time and Embedded Technology and Applications Symposium (RTAS 2005), San Francisco, California, March 7–10, 2005.
11. M. Brownfield, Y. Gupta, and N. Davis IV, "Wireless Sensor Network Denial of Sleep Attack," Systems, Man and Cybernetics (SMC) Information Assurance Workshop, 2005. 15–17 June 2005. Page(s): 356–364.
12. S. Ganeriwal, R. Kumar, and M. B. Srivastava, "Timing-Sync Protocol for Sensor Networks," ACM SenSys 2003, Los Angeles, CA, November 2003.
13. J. Ma, "Mobile Wireless Sensor Network," Advanced Information Networking and Applications Workshops, 2007, AINAW '07. 21st International Conference. Publication Date: 21–23 May 2007, Volume: 2, page(s): 113–120.
14. M. L. Sichitiu and C. Veerarittiphan, "Time Synchronization and Distributed Modulation in Large-Scale Sensor Networks," Electrical and Computer Engineering Department North Carolina State University Raleigh, NC.

15. R. Solis, V. S. Borkar, and P. R. Kumar, "A New Distributed Time Synchronization Protocol for Multihop Wireless Networks," Decision and Control, 2006 45th IEEE Conference. Publication Date: 13–15 Dec. 2006, page(s): 2734–2739.

16. W. Su and I. F. Akyildiz, "Time-Diffusion Synchronization Protocol for Wireless Sensor Networks," EEE/ACM Transactions on Networking (TON), Volume 13, Issue 2 (April 2005). Pages: 384–397.

17. A. A. Syed and J. Heidemann, "Time Synchronization for High Latency Acoustic Networks," INFOCOM 2006. 25th IEEE International Conference on Computer Communications. Proceedings, Publication Date: April 2006, page(s): 1–12.

18. K. Sun, P. Ning, and C. Wang, "Fault-Tolerant Cluster-Wise Clock Synchronization for Wireless Sensor Networks," Transactions on Dependable and Secure Computing, Publication Date: July–Sept. 2005, Volume: 2, Issue: 3, page(s): 177–189.

19. M. Roche, "*Time Synchronization in Wireless Networks*" http://www.cs.wustl.edu/~jain/cse574-06/ftp/time_sync/index.html

20. F. Sivrikaya and B. Yener, "Time Synchronization in Sensor Networks: A Survey," Network, IEEE, Publication Date: July–Aug. 2004, Volume: 18, Issue: 4, page(s): 45–50.

21. J. Elson and D. Estrin, "Time Synchronization for Wireless Sensor Networks," Parallel and Distributed Processing Symposium. Proceedings 15th International, Publication Date: Apr. 2001, page(s): 1965–1970.

22. K. Römer, P. Blum, and L. Meier, "Time Synchronization and Calibration in Wireless Sensor Networks. Handbook of Sensor Networks," Algorithms and Architectures, pp. 199–237, New York: Wiley and Sons, October 2005.

23. J. Elson and K. Römer, "Wireless Sensor Networks: A New Regime for Time Synchronization," ACM Computer Communication. Review (CCR), 33(1):149–154, 2003.

24. D. Ganesan, S. Ratnasamy, H. Wang, and D. Estrin, "Coping with Irregular Spatio-Temporal Sampling in Sensor Networks," SIGCOMM "Computer Communication Review," 34(1):125–130, 2004.

25. B. Liskov, "Practical Uses of Synchronized Clocks in Distributed Systems," in 10th Annual ACM Symposium on Principles of Distributed Computing (PODC' 91), pages 1–10, August 1991.

26. K. Römer, "Temporal Message Ordering in Wireless Sensor Networks," in IFIP Mediterranean Workshop on Ad-Hoc Networks, pages 131–142, June 2003.

27. J. Y. Halpern and I Suzuki, "Clock Synchronization and the Power of Broadcasting," Distributed Computing, 5(2):73–82, 1991.

28. J. Elson and D. Estrin, "Time synchronization for Wireless Sensor Networks (Thesis or Dissertation style)," ACM SIGCOMM Computer Communication Review, Vol. 33 N. 1, p. 149–154.

29. J. Ma, "Mobile Wireless Sensor Network," Advanced Information Networking and Applications Workshops, 2007, AINAW '07. 21st International Conference on, Publication Date: 21-23 May 2007, Volume: 2, page(s): 113–120.

30. J. Elson, and K. Romer, "Wireless Sensor Networks: A New Regime for Time Synchronization," ACM SIGCOMM Comput. Commun. Rev., 2003, 33, (1), pp. 149–154.

31. B. Sundararaman, U. Buy, and A. D. Kshemkalyani, "Clock Synchronization for Wireless Sensor Networks: A Survey," Dept. of Computer Science, Univ. of Illinois at Chicago, March 22, 2005.

32. J. Elson, L. Girod, and D. Estrin, "Fine-Grained Network Time Synchronization Using Reference Broadcasts," Ad Hoc Networks Volume 3, Issue 3, May 2005. Pages 281–323.
33. Q. Li and D. Rus, "Global Clock Synchronization in Sensor Networks," Proc. IEEE Conf. Computer Communications (INFOCOM), Vol. 1, pp. 567–574, Hong Kong, China, Mar. 2004.
34. F. Sivrikaya and B. Yener "Time Synchronization in Sensor Networks: *A Survey,*" IEEE Network, Vol. 18, No. 4, pp. 45–50, Jul.–Aug. 2004.
35. F. Hu, Y. Malkawi, S. Kumar, and Y. Xiao, "Vertical and Horizontal Synchronization Services with Outlier Detection in Underwater Acoustic Networks," Wireless Communications and Mobile Computing, Vol. 8, No. 9, pp. 1165–1181.
36. L. Liu, Y. Xiao, and J. Zhang, "*A Linear Time Synchronization Algorithm for Underwater Wireless Sensor Networks,*" Proc. of IEEE ICC 2009.
37. M. Kopetz, "Global Time in Distributed Real-Time Systems, "Technical Report 15, 1989, Technische Universitat Wien, Wien, Austria.

MEDIUM ACCESS CONTROL

Chapter 7

MAC Protocol Design for Underwater Networks

Challenges and New Directions

Volkan Rodoplu and Amir Aminzadeh Gohari

Contents

7.1 Introduction

In this chapter, we focus on acoustic underwater networks, which utilize acoustic communication links to communicate between the nodes. It is well known that radio frequency (RF) signals do not propagate well underwater, and the optical signals [1] are useful for short distances and under clear water propagation conditions. For long distances, the Navy has utilized acoustic communication. More recently, there has been a growing emphasis in the research community on underwater ad hoc and sensor networks [2–6], which utilize acoustic communication over relatively short distances of about 100 meters. The advantages of such short distance communication are numerous: First, the strong multipath profile [7] that exists in the traditional

179

applications over much longer distances is significantly reduced, thereby increasing the data rate at which the communication can take place. Second, the propagation delays are reduced. This has important consequences for the design of medium access control (MAC) protocols because achieving a rough synchronization between the nodes becomes possible. Third, short distances save battery energy [8], which is especially important for underwater sensor networks. For example, in scientific data collection applications [9], in which the batteries are expected to last for months under difficult conditions, the battery energy becomes the most important constraint. In contrast with terrestrial communications, underwater acoustic communications is affected by both distance-dependent path loss, as well as frequency-dependent absorption [10]. On the dB scale, the path loss increases logarithmically with distance, whereas the frequency-dependent absorption increases linearly with distance. Hence, over long distances, the latter dominates, and leads to significant energy consumption. By operating sensor networks with internodal distances of about 100 meters, the frequency-dependent absorption term becomes only a small fraction of the path loss, hence its deleterious effects on energy can be significantly reduced.

This chapter focuses on the design of MAC protocols for underwater sensor networks. The recent attention to this topic is motivated by several factors: First, in the marine science, limnology, and oceanography communities, there has been a growing need for underwater sensor networks. Currently, many of the underwater sensors in Long-Term Ecological Research (LTER) projects [9] are stand-alone. The future vision for these projects is for the sensors to be networked with each other such that (1) adaptive sampling becomes possible when sensors in a region detect an interesting event so that more samples can be taken on demand, (2) in-network data aggregation and fusion become possible, and (3) the collected data can be sent to the laboratory in a timely fashion to lead to quick analysis. There is a great thrust [11] toward the collection and visualization of real-time data in these scientific communities, such that the lakes and oceans can be observed on a continuous basis. The design of MAC protocols for these underwater sensor networks will enable effective data collection, aggregation, and networking to the sites where the data can be displayed and interpreted.

Second, MAC protocols for underwater networks are important for the Navy. Much of the research in the last 30 years on underwater networks has focused not on networking, but rather on point-to-point communications. Indeed, very sophisticated, successful methods [12–14] have been designed for the physical layer, with superior performance. The underwater acoustic physical channel is one of the severest out of all of the channels found in nature: The Doppler and multipath profiles, as well as the strong distance-dependent loss leaves very little effective bandwidth over which to communicate. Hence, well-designed equalizers [15] have been the key to achieving the capacity of these channels. In contrast with the physical layer developments, however, the MAC protocols used for the Navy applications have largely been based on the CSMA/CA Carrier Sense Multiple Access with Collision Avoidance prototype. The RTS/CTS exchange is in fact not very well suited to underwater communication due to the propagation delay [16]. Hence, both the hidden and exposed terminal problems are exacerbated. One of the reasons that it has been difficult to assess MAC protocols for underwater networks is the lack of a measuring stick that is comparable to Shannon capacity. For the physical layer, the researchers are able to assess the gap from Shannon capacity, whereas no similar MAC protocol capacity measures have been developed for underwater networks. For terrestrial networks, the g-put is a measure of the channel utilization; however, when long propagation delays are involved, as in underwater acoustic networks, it is no longer possible to attribute a single utilization measure to the channel itself. Instead, a utilization measure must be attributed to each individual node's receiver. Hence, a part of the research on MAC protocol design for underwater networks today focuses on defining appropriate capacity measures for underwater MAC protocols. A second part of the effort focuses on the development of protocols [17–24].

The rest of this chapter is organized as follows: Each of the main sections of this chapter is devoted to a research challenge and outlines new directions to address the research challenge. In Sections 7.2, 7.3, and 7.4, the research challenges are the propagation delay of the acoustic medium, energy consumption, and mobility, respectively. Finally, in Section 7.5, we present our conclusions.

7.2 Challenge I: Propagation Delay

Large and location-dependent delays in underwater networks degrade the performance of traditional medium access protocols [5,25]. Therefore, the estimation of propagation delay can be employed to improve the performance of MAC protocols for such networks [26,27] or to achieve synchronization between the nodes. The challenge that long propagation delays pose for the design of MAC protocols can be addressed in a couple of ways: First, the propagation delay can be estimated by a handshake between the nodes, as shown in Figure 7.1. Node A, which wants to estimate the propagation delay to its neighbor Node B, sends a beacon signal and records when the beacon signal was sent out. When Node B receives the beacon signal, it immediately returns an acknowledgment (ACK) to Node A, which then records when the ACK was received. Assuming that the propagation delay is symmetric between A and B, Node A divides by two, the difference between the time that the beacon was sent and the time that the ACK was received, to obtain an estimate of the propagation delay. A three-way handshake can be performed to allow the estimation of the propagation delay by Node B, as well. When Node A receives the ACK, it returns an ACK to Node B right away. Based on when its own ACK was sent out and the ACK from Node A was received, Node B itself can also estimate the propagation delay by a similar procedure. These calculations assume that the processing delay at any receiving node is small, compared to the propagation delay, which is almost always satisfied since the processors use silicon technology and operate at speeds much faster than those in the acoustic medium.

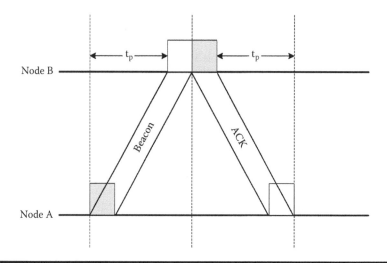

Figure 7.1 Simple handshake algorithm for propagation delay estimation between two nodes.

There are variations on this simple propagation delay estimation scheme, which are also possible. For example, as shown in Figure 7.2, when Node B receives the beacon signal from Node A, it may choose to record the time it was received, record the time until its own ACK is sent out, and place a time stamp in its own ACK that indicates the time that passed from the time it received A's beacon signal to the time that it sent out the ACK. This may be useful, for example, if Node B is not able to send back an ACK right away because it is busy communicating to another node, say Node C. When Node A receives the ACK from Node B, under this estimation scheme, it first decodes the stamp in the ACK signal, then subtracts this delay that was incurred at Node B before it divides by two in order to obtain an estimate of

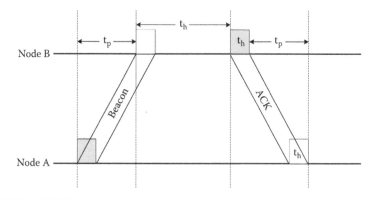

Figure 7.2 Propagation delay estimation sending delayed ACK frames.

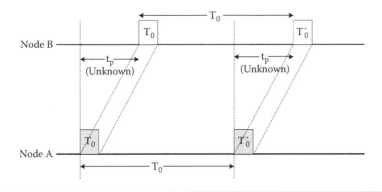

Figure 7.3 Relative wake up time advertisement. (After M. K. Park and V. Rodoplu, *IEEE J. Oceanic Engineering,* vol. 32, no. 3, pp. 710–720, July 2007.)

the propagation delay. Both references [20] and [28] utilize this method as part of the MAC protocol design.

The second way to address the challenge of propagation delay in underwater acoustic networks is to use relative time stamps in the beacon signals, and not estimate the propagation delay at all. In this method, which is shown in Figure 7.3, each node sends out a beacon signal, advertising to its neighbors exactly when it will send again, as a time difference from the current transmission, according to its own clock. Let T_0 be the time on this time stamp. Hence, Node A states in its beacon signal that it will send again T_0 seconds later. When a neighbor, say Node B, receives this announcement a random and unknown propagation delay later, it sets its clock so that it will be ready to receive the transmission from Node A, T_0 seconds later, according to Node B's clock. If the propagation delay does not change significantly and if the clock drift is not significant within T_0 seconds, then Node B indeed receives the next transmission from Node A exactly when it expects it. Note that the propagation delay need not be estimated at all. Hence, the nodes achieve pseudo-synchronization without any estimation of the propagation delay.

The specialization of this method to periodic announcements appeared in the literature in [8] in the context of topology control for energy-limited, terrestrial networks, and in [29] in the context of MAC protocols for terrestrial, energy-limited networks. For underwater networks, this was utilized in the underwater acoustic network (UWAN)-MAC protocol [17,18]. However, note that the announcement need not be periodic. In fact, at any slot where a new announcement to the neighbors is made, a new time T_0 can be specified. In UWAN-MAC, this is covered under the context of changing the period of the transmissions; however, no periodicity of successive transmissions is necessary.

There are both advantages and disadvantages of the first and second methods described above. By virtue of the first method, in which the propagation delay is estimated directly by a three-way handshake, the knowledge of the propagation

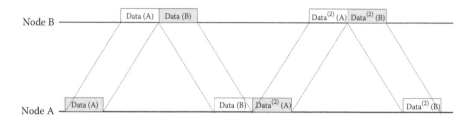

Figure 7.4   Traditional stop and wait transmission protocol.

delay allows the nodes later in the DATA transmission phase to set timers for the ACK messages. If an ACK for a DATA packet has not been received after a time-out plus twice the propagation delay, then a retransmission becomes possible at the data link layer. Hence, the estimation of the propagation delay allows the protocols to wait an extra duration of twice the propagation delay and effectively turns the internodal channel into one that does not involve propagation delays. However, an inefficiency due to the extra wait time of propagation delay is incurred as a result. In a medium that is shared by all the nodes in a network, even a TDMA (Time Division Multiple Access)-based scheme that allocates slots to different nodes using a guard time that is at least as large as the propagation delay can be set up, after the propagation delay has been estimated. Hence, "slotted protocols," as a general class, as described in [26] and [30], become possible; however, although such MAC protocols that aim to turn the underwater channel into an RF-like channel, with an efficiency cost incurred due to these guard times, are useful for the short distances we consider, they cannot achieve the fundamental limits of MAC utilization at each receiver. In order to understand this, consider the example of two nodes that share the same medium, and which need to send information to each other on a periodic basis. In the two-node case, no explicit estimation of the propagation delay is required, and the nodes can send the DATA to each other in a handshaking fashion. Figure 7.4 illustrates such a traditional scheme.

The utilization on each node's axis is

$$\frac{t_{DATA}}{t_{DATA}+t_p}$$

where t_{DATA} is the DATA duration, and t_p is the propagation delay. Hence, only in the limit as t_{DATA} goes to infinity can the utilization approach 100% (not taking into account any data link layer errors here, and counting only MAC utilization). Now, if the value of the propagation delay is known, then the transmission shown in Figure 7.5 becomes possible, where each node transmits data only for a duration of t_p, and then receives data for the next duration of t_p. We call this transmission "butterfly transmission."

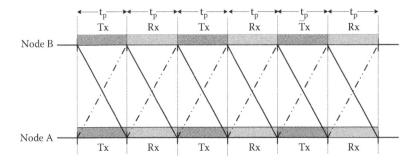

Figure 7.5 **Butterfly transmission to maximize throughput.**

Butterfly transmission has 100% utilization on each of the nodes' axes. It is fair, in the sense that it allows each node to transmit 50% of the time and receive 50% of the time; hence, both nodes can send their data in a timely fashion. This procedure requires perfect synchronization between the nodes so they know that they can start the butterfly transmission at the same time, on a global axis. This cannot be achieved with 100% accuracy in all circumstances; however, the following describes a case where it can: Assume that one of the nodes is designated *a priori*, as a "leader," and the other one always as a "follower." The follower never initiates protocol control information exchanges itself, and a leader always does. This designation is hardwired into these two nodes upon deployment. Then, as shown in Figure 7.6, the leader, Node A, initiates a three-way handshake for the estimation of propagation delay. Then, the leader starts its DATA transmission, in packets of duration t_p, exactly t_p seconds after its last transmission, in the three-way handshake. The follower, Node B, also estimates t_p by the end of the three-way handshake, and immediately starts transmitting its own data, in packets of duration t_p, right after the end of the three-way handshake. The processing delays are very small compared with propagation delays, and are hence ignored in the figure. Hence, the

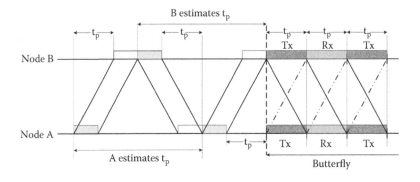

Figure 7.6 **Leader-follower three-way handshake to start butterfly transmission.**

nodes can settle into the butterfly transmission with 100% utilization after this three-way handshake.

The leader in the above setup was hardwired into the nodes. The leader election problem, without such a hardwiring, is not trivial, even in the two-node case. The main complication arises from the fact that without a predetermined leader, the nodes can send at any time to try to establish the leader; however, the possibility of collisions with a positive probability always leaves some chance that synchronization is not achieved. Hence, an interesting area of research is understanding how such leader election problems, which are fundamental in computer science, can be addressed in media with variable propagation delay, as in underwater networks. Many of the algorithms for leader election in the literature that aim to minimize the message overhead on a topology, however, assume that the links can exchange information without collisions.

The butterfly scheme, which we developed for the two-node case, cannot be generalized to the three-node case, that is, when three nodes share the same MAC medium, with variable propagation delays. In fact, for the three-node case, there are always scenarios where 100% utilization on every node's time axis is not possible. In the literature, [31] sheds considerable light on the optimal MAC scheduling problem (albeit with minimum energy as the main objective). An important contribution of that paper is the formulation of the scheduling and routing problem with variable propagation delays. This is formulated as an optimization program, and optimal MAC schedules can be generated for any placement of the nodes. The paper goes further and derives the optimal placement of the nodes to achieve the minimum energy solution under interference constraints. Hence, the results of that paper give a general scheduling and routing framework to achieve the lowest energy consumption possible in an underwater network with variable propagation delays. However, the downside is that the solution is a scheduling solution; that is, it is a centralized solution that assumes that all of the propagation delays are known *a priori* (or have been estimated), and a centralized scheduler tells the nodes, on side information channels, exactly when to start and end their transmissions. In reality, such side information channels are difficult to establish, especially in the presence of propagation delays. In a framework of MAC protocol capacity, what is needed is an achievable upper bound that can show what the maximum utilization can be at each node's time axis, *given* that the propagation delay information is not *a priori* available, and taking into account the overhead that would be incurred by having to transmit that control information. The difficulty of such a formulation is not particular to underwater networks, and is an unsolved problem in both communications and networking. Currently, there are two traditional approaches: A global scheduler is assumed to derive the capacity of a globally scheduled solution, and then ad hoc methods are utilized to distribute the required side information on control channels. This is invariably the approach of the communications community. On the networking side, notions of protocol optimality under side information assumptions are largely missing; hence, protocols in general are designed in an ad hoc fashion, and the goodness of a protocol is understood only in comparison with other competing proposals, and their relative performance evaluation in rather specialized circumstances. What is actually needed is a general framework that can precisely define protocol optimality, and also describe methods of how to derive or generate such optimal protocols. These would need to go beyond scheduling solutions that do not make the overhead of the control information as part of the capacity formulation.

7.3 Challenge II: Energy Consumption

A second challenge that the design of underwater sensor networks has to face is energy. The limited battery energy of the devices makes it paramount to design solutions that will allow the nodes to operate for months without having to recharge or replace their batteries [5,32]. The challenges here are not entirely different from those present in the design of terrestrial sensor networks. The most common solution, from the perspective of MAC layer design, is the use of sleep schedules, as originally proposed in S-MAC (Senser Medium Access Control) [29]. For underwater networks, UWAN-MAC [18] certainly adopts this solution; however, with the recognition that the propagation delays are large and unknown, it opts for a solution that uses randomly selected transmit times. Unlike S-MAC, no attempt is made to endeavor to make the schedules of clusters to converge to some global schedule, which cannot be attained with variable propagation delays. In addition, RTS/CTS exchanges are completely avoided in UWAN-MAC, in order not to incur any energy overhead. Indeed, the figures of merit in this context are (1) total energy consumed, (2) fraction of transmit energy consumed due to collisions, and (3) fraction of receive energy consumed due to collisions. These metrics are entirely different from those used to evaluate ALOHA, MACA (Multiple Access Collision Avoidance), and MACAW (Multiple Access Collision Avoidance for Wireless), namely the traditional protocols, which assume that bandwidth is the limiting resource rather than energy. When the amount of data that is collected from the sensors is small, then the protocols must become cognizant of this fact; hence, protocol overhead can no longer be made to go to zero in percentage by making the DATA durations very large, since large amounts of data are simply not available. It is clearly seen that protocol design has to be entirely different in regimes where data generation rates are high, and where they are low. Currently, there is no "protocol engine" that is able to generate protocols on demand, based on which regime is in question. Even an understanding of which protocols are optimal in which regimes is largely missing. If formulated, those would be important contributions to making significant progress in the design of MAC protocols for underwater networks.

The need for protocol generation, or at least configurability, is also clearly seen when the modem specs for acoustic modems are considered. In [33], modem specs for an air modem and an acoustic modem are presented, and are seen to be widely different in terms of sleep, idle, receive, and transmit power consumptions. Even within the domain of acoustic modems, there is wide variability. Protocols, by their nature, are designed with certain assumptions in mind on modem specifications. For example, UWAN-MAC explicitly assumes that the power consumption in the sleep mode is much lower than that in the idle mode. Hence, there is an incentive to turn off the node as much as possible, just as is assumed in S-MAC. However, if the sleep and idle power consumptions are comparable for a particular modem that is chosen, then UWAN-MAC protocol becomes far from optimal. A good protocol for such a modem choice would eliminate the sleep schedules, and would keep the node on, in order to minimize the delay. Hence, ideally, what is needed is the generation of a protocol on-demand, based on a variety of variables including the modem specifications, as well as the application requirements (e.g., QoS constraints on delay, energy, and bandwidth). The software-defined radios cannot address this challenge: Even though they provide configurability of parameters at the software level, there is no methodology to drive the protocols toward solutions that are optimal for particular modem specs, and application constraints. In terms of the available optimization engines, we now know very well how to do numerical optimization; that is, setting of the optimal values of particular, pre-determined parameters, but not protocol optimization, where we can embed structurally different protocols in a protocol space and efficiently iterate over those to find the optimal protocol. New mathematical methods are necessary to address this challenge, and their impact on networking research would be profound.

Energy, as a valuable resource, has recently led researchers to look at network capacity in a fundamentally different way. As early as the 1980s, the bits-per-joule capacity of a point-to-point underwater link [34] was defined and evaluated. More recently, for terrestrial networks, the bits-per-joule capacity of an energy-limited wireless network [35,36] was defined as the maximum number of bits that can be transferred reliably between sources and destinations divided by the total number of joules of energy deployed into the network. Here, reliability is different from Shannon reliability and presumes only that a fixed, but arbitrarily small probability of error ε is achieved end-to-end, for each transfer. The bits-per-joule capacity is a suitable measure of capacity for networks with energy-limited devices, which aim to transfer delay-insensitive traffic to their destinations. In the context of scientific data collection, this is often the case, as the delay constraint required for the bits to reach their destination is on the order of minutes rather than milliseconds, as would be for real-time applications. The bits-per-joule capacity concept is equally applicable to underwater sensor networks, which are made up of small, energy-limited sensor nodes that aim to transfer delay-insensitive data to a collection site. The main difference is in the channel model where, in addition to path loss, a frequency-dependent absorption term is present. However, the scaling law for the bits-per-joule capacity of an underwater network is similar to the one for a terrestrial one, since the frequency-dependent absorption term $a(f)^{-d}$ goes to 1 as the internodal distance d goes to 0. Here, $a(f)$ is the absorption coefficient as a function of carrier frequency. The design of MAC protocols for energy-limited, underwater networks can focus on the maximization of the bits-per-joule performance of these

Figure 7.7   Mobile node B moves away from stationary node A at a constant velocity v_B.

networks, rather than throughput, as would be the case in traditional, bandwidth-limited application domains. Again, in terms of the protocol generation problem, it is possible to pose questions such as, what is the MAC protocol that maximizes the bits-per-joule performance? The development of mathematical tools that will allow the formulation and solution of such a protocol design optimization problem would indeed further the research on MAC protocol design for underwater networks.

7.4 Challenge III: Mobility

Many underwater sensor networks are made up of only stationary nodes; however, increasingly, mobility is becoming an important aspect of the design of MAC protocols for the underwater medium [5,6]. As early as in Sea Web in the period 1998 through 2000 [37], autonomous underwater vehicles (AUVs) have been an important part of the underwater network landscape. The MAC layer of Sea Web was based on the traditional CSMA/CA solution, and many of the proposed protocols [38,39] for underwater networks since then have been based on CSMA/CA. However, mobility itself introduces new challenges and a call for new models in the design of MAC protocols. We illustrate this with a few examples: First, consider the example of two nodes in Figure 7.7. In this figure, Node A is stationary, and Node B is moving away from Node A, at a constant velocity v_B. Figure 7.8 shows what the butterfly transmission solution for Figure 7.7 would look like in this case.

Because the propagation delay t_p is getting larger, the intervals of duration t_p on the diagram are getting larger as well. The 100% utilization of both of the time axes of the nodes is maintained, as well as the fairness that gives equal opportunities to

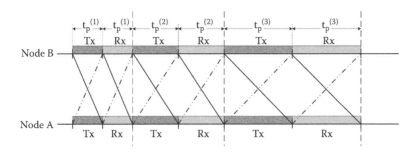

Figure 7.8   Butterfly transmission solution for the case of Figure 7.7.

both nodes to transmit and to receive. The shown solution is possible if and only if (1) one of the nodes is a designated leader in initiating the transmission, to be able to set up this transmission in the first place, and (2) the propagation delay, as it changes, can be perfectly estimated. For example, if Node B's velocity is known, this can be used to determine how the duration of each transmit interval on the diagram should be increased upon successive transmissions.

This brings us to one of the important general lessons in the choice of whether the propagation delay should be estimated or not. In analogy with the coherence time of an impulse response at the physical layer, we can define the "coherence time of propagation delay" at the network layer. For transmission between two completely stationary nodes (i.e., such that even platform motion around a fixed position is not possible), the coherence time of the propagation delay is infinite. In this case, the propagation delay can be estimated once and for all, and 100% utilization on each of the nodes' time axes can ensue. However, if the coherence time of the propagation delay is bounded, then the estimation of the propagation delay makes sense only if the coherence time is much larger than the time required to estimate the propagation delay. This bears an analogy to the physical layer theorem that if the coherence time of the impulse response is much larger than the symbol duration, then the physical layer channel can be estimated. However, the difference in the resolution required between the estimation of the channel tap coefficients at the physical layer and the estimation of the propagation delay at the MAC layer via a handshake should be carefully noted. The physical layer estimation must happen much faster, and the resolution required is at the level of precision of the impulse response coefficients. From the perspective of the MAC layer, the underlying physical layer coefficients can change. As long as this does not have an appreciable effect on the propagation delay experienced, the coherence time required for propagation delay estimation at the MAC layer can be much larger than the coherence time of the impulse response of the physical layer channel. For example, consider the case of two nodes whose positions randomly move about their centers (also known as platform motion, referring to the random motion of the platform on which the node is located, e.g., due to water currents). Then, the propagation delay, averaged over these small spatial movements around a central position, remains the same, whereas the impulse response coefficients at the physical layer can change dramatically. Hence, the coherence time required at the MAC layer is much larger.

An open question is the derivation of the MAC protocol capacity, for a set of N nodes, each of which may be mobile or stationary. Here, by MAC protocol capacity, we mean the maximum utilization possible at each of the nodes' time axes. This will necessarily be defined as a region of possible utilizations in an N-dimensional space, rather than a single number that we can call capacity. The example in Figure 7.9 illustrates some of the difficulties in designing MAC protocols that truly exploit the mobility of some of the nodes. In this figure, Node B and Node C are stationary nodes, and Node A is an AUV that moves between B and C. In this case, if Node A has something important to broadcast to both B and C, how can it interrupt an ongoing transmission

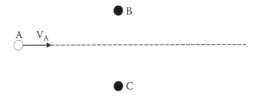

Figure 7.9   Mobile AUV moves toward stationary nodes B and C.

between B and C, and broadcast its data in the little time that it has in passing between the two nodes? Several different strategies might be possible: (1) Stationary nodes may always leave certain slots in their transmissions empty, for possible usage by passing mobile nodes. (2) A separate frequency can be allocated as a control channel, on which a newcomer passerby can advertise its presence. When this beacon is heard, the stationary nodes immediately pause their transmissions to listen to what the passerby has to broadcast. (3) A CDMA (Code Division Multiple Access)-based solution is pursued in which certain codes are allocated as channels for possible passersby [40,41]. When those channels are utilized, the ongoing transmission between B and C, which uses pseudonoise sequences, is degraded softly, while the new transmission from the passerby can be detected and decoded on the parallel channel. All of this discussion shows that MAC protocols should take mobility directly into account, and utilize mobility in order to achieve the highest utilization at each node. The estimation of the propagation delays for a large set of nodes may become prohibitive, especially when a large number of *mobile* nodes are involved. Hence, a more aggregate view [42] of such networks, and the characterization of the propagation delay as a function over space rather than between node pairs, may become more useful when very large-scale mobile networks are in question.

7.5 Conclusions

In summary, the following would benefit greatly the design of MAC protocols for acoustic underwater networks:

1. A definition of MAC protocol capacity that is cognizant of the propagation delays of the underwater acoustic medium: In contrast with RF communication, no single utilization metric can be ascribed to the channel itself. Instead, utilization must be measured along each node's own time axis. The MAC protocol capacity can be defined as the set of Pareto-optimal points of an N-dimensional set of feasible utilization rates of each of the N nodes. Such a definition would provide a method to compare the performance of different protocols, and to measure how far from the nearest Pareto-optimal solution each protocol is.

2. A methodology for the generation of MAC protocols on demand. Such a protocol generator would generate the optimal protocol, given (A) the

physical layer resource parameters, such as the transmit, receive, idle, and sleep power specifications of the modem used; (*B*) the channel model; and (*C*) the application-layer requirements, such as the required QoS. The design of such a protocol generator is far from trivial; however, its impact on both underwater networking and networking at large would be profound.

3. A definition of "protocol optimality." For example, when does it make sense to estimate the propagation delay? A MAC protocol that estimates the propagation delay on a periodic basis and uses this for scheduling decisions can be said to be a "coherent MAC protocol," in analogy with coherent physical layer communication. A MAC protocol that does without the estimation of propagation delays can be said to be "non-coherent." If we divide the class of MAC protocols into coherent and non-coherent ones, how do we determine an optimal protocol within each class? Finally, how do we determine the optimal MAC protocol overall that also involves the choice of whether or not to estimate the propagation delay, based on the cost incurred? Hence, the problem of the determination of an optimal protocol can be decomposed into a choice between several subproblems, each of which makes different assumptions of the side information available (e.g., coherent vs. non-coherent). Once an optimal protocol within each class is determined, these protocols can then be compared by examining the extra cost incurred in providing the side information that each class requires, and finally comparing the performance of each protocol, now with this overall total cost in mind.

4. The incorporation of mobility models into the design of MAC protocols. In this chapter, we have already discussed the coherence time of propagation delay as a novel MAC-layer concept, and both the impact and exploitation of mobility for MAC protocol design. Both link-level and aggregate network-layer coherence time concepts can be defined for underwater networks, with the goal of determining optimal MAC protocols for large-scale networks that involve both mobile and stationary nodes.

References

1. I. Vasilescu, K. Kotay, D. Rus, M. Dunbabin, and P. Corke, "Data collection, storage, and retrieval with an underwater sensor network," in *Proc. of the 3rd international Conference on Embedded Networked Sensor Systems (Sen-Sys'05)*, pp. 154–165, Nov. 2005.
2. I. F. Akyildiz, D. Pompili, and T. Melodia, "Challenges for efficient communication in underwater acoustic sensor networks," *IEEE/ACM Sighed Review*, vol. 1, no. 2, pp. 3–8, July 2004.
3. I. F. Akyildiz, D. Pompili, and T. Melodia, "Underwater acoustic sensor networks: research challenges," *Ad Hoc Networks Journal*, vol. 4, no. 3, pp. 257–279, May 2005.
4. J. Heidemann, W. Ye, J. Wills, A. Syed, and Y. Li, "Research challenges and applications for underwater sensor networking," in *Proc. of the IEEE Wireless Communications and Networking Conference*, pp. 228–235, Apr. 2006.

5. J. Partan, J. Kurose, and B. N. Levine, "A survey of practical issues in underwater networks," in *Proc. of the 1st ACM Workshop on Underwater Networks (WUWNet'06)*, pp. 17–24, Sep. 2006.

6. I. F. Akyildiz, D. Pompili, and T. Melodia, "State-of-the-art in protocol research for underwater acoustic sensor networks," in *Proc. of the 1st ACM Workshop on Underwater Networks (WUWNet'06)*, pp. 7–16, Sep. 2006.

7. H. V. Poor and G. W. Wornell, *Wireless Communications*. NJ: Prentice Hall, 1998, Chapter 8. Underwater acoustic communications.

8. V. Rodoplu and T. H. Meng, "Minimum energy mobile wireless networks," *IEEE Journal on Selected Areas in Communications*, vol. 17, pp. 1333–1344, Aug. 1999.

9. "Lake Metabolism Project," http://www.lakemetabolism.org

10. M. Stojanovic, "On the relationship between capacity and distance in an underwater acoustic communication channel," in *Proc. of the 1st ACM Workshop on Underwater Networks (WUWNet'06)*, pp. 41–47, Sep. 2006.

11. "Global Lake Ecological Observatory Network (GLEON)," http://gleon.org

12. M. Stojanovic, "Recent advances in high-speed underwater acoustic communications," *IEEE J. Oceanic Eng.*, vol. 21, pp. 125–136, 1996.

13. T. C. Yang, "Correlation-based decision-feedback equalizer for underwater acoustic communications," *IEEE J. Oceanic Eng.*, vol. 30, pp. 865–880, 2005.

14. J. C. Preisig, "Performance analysis of adaptive equalization for coherent acoustic communications in the time-varying ocean environment," *Journal of the Acoustical Society of America*, vol. 118, no. 263, 2005.

15. M. Stojanovic, J. Proakis, and J. Catipovic, "Performance of high-rate adaptive equalization on a shallow water acoustic channel," *Journal of the Acoustical Society of America*, vol. 100, no. 4, pp. 2213–2219, 1996.

16. P. Xie and J. Cui, "Exploring random access and handshaking techniques in large-scale underwater wireless acoustic sensor networks," in *Proc. of the MTS/IEEE OCEANS 2006*, pp. 1–6, Sep. 2006.

17. V. Rodoplu and M. K. Park, "An energy-efficient MAC protocol for underwater wireless acoustic networks," in *Proc. of the MTS/IEEE OCEANS 2005*, pp. 1198–1203, Sep. 2005.

18. M. K. Park and V. Rodoplu, "UWAN-MAC: An energy-efficient MAC protocol for underwater acoustic wireless networks," *IEEE J. Oceanic Engineering*, vol. 32, no. 3, pp. 710–720, July 2007.

19. P. Casari, F. E. Lapiccirella, and M. Zorzi, "A detailed simulation study of the UWAN-MAC protocol for underwater acoustic networks," in *Proc. of MTS/IEEE Oceans 2007*, Sep. 2007.

20. X. Guo, M. R. Frater, and M. J. Ryan, "A propagation-delay-tolerant collision avoidance protocol for underwater acoustic sensor networks," in *Proc. of the MTS/IEEE OCEANS 2007-Asia Pacific*, pp. 1–6, May 2007.

21. J. K. Yeo, Y. K. Lim, and H. H. Lee, "Modified MAC (media access control) protocol design for the acoustic-based underwater digital data communication," in *Proc. IEEE International Symposium on Industrial Electronics (ISIE'01)*, vol. 1, pp. 364–368, 2001.

22. M. Molins and M. Stojanovic, "Slotted FAMA: A MAC protocol for underwater acoustic networks," in *Proc. of the MTS/IEEE OCEANS 2006-Asia Pacific*, pp. 1–7, May 2006.

23. B. Peleato and M. Stojanovic, "A MAC protocol for ad-hoc underwater acoustic sensor networks," in *Proc. of the 1st ACM Workshop on Underwater Networks (WUWNet'06)*, pp. 113–115, Sep. 2006.

24. H. Tan and W. K. G. Seah, "Distributed CDMA-based MAC protocol for underwater sensor networks," in *Proc. of the 32nd IEEE Conference on Local Computer Networks (LCN'07)*, pp. 26–36, Oct. 2007.

25. A. A. Syed, W. Ye, B. Krishnamachari, and J. Heidemann, "Understanding spatio-temporal uncertainty in medium access with ALOHA protocols," in *Proc. of the 2nd ACM Workshop on Underwater Networks (WUWNet'07)*, pp. 41–48, Sep. 2007.

26. N. Chirdchoo, W.-S. Soh, and K. C. Chua, "ALOHA-based MAC protocols with collision avoidance for underwater acoustic networks," in *Proc. of 26th IEEE International Conference on Computer Communications (INFOCOM'07)*, pp. 2271–2275, May 2007.

27. Y. Chen and H. Wang, "Ordered CSMA: A collision-free MAC protocol for underwater acoustic networks," in *Proc. of MTS/IEEE Oceans 2007*, Sep. 2007.

28. P. Xie and J. Cui, "R-MAC: an energy-efficient MAC protocol for underwater sensor networks," in *Proc. of International Conference on Wireless Algorithms, Systems and Applications (WASA'07)*, pp. 187–198, Aug. 2007.

29. W. Ye, J. Heidemann, and D. Estrin, "An energy-efficient MAC protocol for wireless sensor networks," in *Proc. of the 21th International Annual Joint Conference of the IEEE Computer and Communications Societies (INFOCOM 2002)*, vol. 3, pp. 1567–1576, Jun. 2002.

30. K. B. Kredo II and P. Mohapatra, "A hybrid medium access control protocol for underwater wireless networks," in *Proc. of the 2nd ACM Workshop on Underwater Networks (WUWNet'07)*, pp. 33–40, Sep. 2007.

31. L. Badia, M. Mastrogiovanni, C. Petrioli, S. Stefanakos, and M. Zorzi, "An optimization framework for joint sensor deployment, link scheduling and routing in underwater sensor networks," in *Proc. of the 1st ACM Workshop on Underwater Networks (WUWNet'06)*, pp. 56–63, Sep. 2006.

32. A. Porto and M. Stojanovic, "Optimizing the transmission range in an underwater acoustic network," in *Proc. of MTS/IEEE Oceans 2007*, Sep. 2007.

33. A. F. Harris III, M. Stojanovic, and M. Zorzi, "When underwater acoustic nodes should sleep with one eye open: Idle-time power management in underwater sensor networks," in *Proc. of the 1st ACM Workshop on Underwater Networks (WUWNet'06)*, pp. 105–108, Sep. 2006.

34. H. M. Kwon and T. G. Birdsall, "Channel capacity in bits per joule," in *IEEE J. Oceanic Engineering*, vol. 11, no. 1, pp. 97–99, Jan. 1986.

35. V. Rodoplu and T. H. Meng, "Bits-per-joule capacity of energy-limited wireless ad hoc networks," in *Proc. of IEEE GLOBECOM*, vol. 1, pp. 16–20, Nov. 2002.

36. V. Rodoplu and T. H. Meng, "Bits-per-joule capacity of energy-limited wireless networks," *IEEE Transactions on Wireless Communications*, vol. 6, no. 3, pp. 857–865, Mar. 2007.

37. J. Rice, B. Creber, C. Fletcher, P. Baxley, K. Rogers, K. McDonald, D. Rees, M. Wolf, S. Merriam, R. Mehio, J. Proakis, K. Scussel, D. Porta, J. Baker, J. Hardiman, and D. Green, "Evolution of Sea Web underwater acoustic networking," in *Proc. of MTS/IEEE Oceans 2000*, vol. 3, pp. 2007–2017, Sep. 2000.

38. M. Tao, S. Haoshan, and W. Yu, "A MAC protocol for underwater sensor networks," in *Proc. of 8th International Conference on Electronic Measurement and Instruments (ICEMI'07)*, pp. 144–148, Aug. 2007.

39. B. Creber, J. Rice, P. Baxley, and C. Fletcher, "Performance of undersea acoustic networking using RTS/CTS handshaking and ARQ retransmission," in *Proc. of MTS/IEEE Oceans 2001*, vol. 4, pp. 2083-2086, Nov. 2001.

40. D. Pompili, T. Melodia, and I. F. Akyildiz, "A distributed CDMA medium access control for underwater acoustic sensor networks," in *Proc. of Mediterranean Ad Hoc Networking Workshop (Med-Hoc-Net)*, June 2007.

41. F. Salva-Garau and M. Stojanovic, "Multi-cluster protocol for ad hoc mobile underwater acoustic networks," in *Proc. of the IEEE OCEANS 2003*, vol. 1, pp. 91–98, Sep. 2003.

42. M. K. Park and V. Rodoplu, "Energy maps for large-scale, mobile wireless networks," in *Proceedings of the IEEE International Communications Conference (ICC 2007)*, June. 2007.

Chapter 8

Dynamic TDMA- and MACA-Based Protocols for Distributed Topology Underwater Acoustic Networks

Shiraz Shahabudeen, Mandar Chitre, and Mehul Motani

Contents

In this chapter, we explore the design choices for medium access control (MAC) in an underwater acoustic network (UAN) primarily for a distributed topology. Though there are differences such as high latency in UANs as compared to terrestrial networks, the basis for network protocol design is very similar in both types of networks. In order to provide a strong basis for protocol type selection, the chapter first provides insights into the relative performance of traditional multiple access protocols— code division multiple access (CDMA), time division multiple access (TDMA), and frequency division multiple access (FDMA)—and illustrates why, in a distributed topology, CDMA and FDMA require full-duplex and multi-channel functionality to have similar performance as TDMA. We also evaluate the often-cited capacity advantage of CDMA over FDMA and TDMA in cellular networks with re-use, and illustrate the advantages of dynamic forms of TDMA in distributed ad hoc networks. Medium access with collision avoidance (MACA)-based protocols are shown to be closely related to the dynamic TDMA protocol, and they address the UAN MAC problem very well. We also provide some analytical performance estimates for a MACA-based protocol, for easy assessment of the effect of parameters such as latency, bit error rate (BER), packet detection probability, etc., in a UAN and show that the use of packet trains improves network efficiency.

8.1 Introduction

A key problem that any UAN has to address is that of MAC—how do multiple nodes coordinate access to the acoustic channel? In this chapter, we evaluate various possible MAC options and aim to provide useful insights into the selection of MAC for UANs.

The MAC problem for UANs is conceptually similar to the MAC problem for terrestrial radio wireless networks. However, there are some key differences between the two problems: larger propagation delay due to low sound speed, extremely low point-to-point data rates, and high raw BER. Thus researchers often suggest that MAC protocols for UANs should be developed from the ground up and not directly adopted from existing terrestrial protocols.[1,2]

The overall networking problem includes MAC, multi-hop routing, reliability, data transfers to the wider Internet infrastructure, etc. In this chapter, we focus only on the MAC layer. Two of the most relevant topologies for UANs are distributed and centralized. There are other specialized topologies such as linear, ring, etc., that we do not consider, as they are only applicable in special situations in UANs. In order to avoid collisions, we can use code, frequency, or time division to separate logical communication channels. This leads to the common multiple access schemes—CDMA, FDMA, and TDMA. Space division multiple access (SDMA) is rarely used as it is impractical to implement it except in special scenarios. We define *channelization* as the process of dividing the total available channel capacity in the acoustic channel into a set of logical *channels* for the purposes of multiple access and spatial re-use. Channelization can either be static or dynamic. Spatial re-use of channels is important in UANs since they are both power- and bandwidth-limited and consequently capacity-limited. UANs could also employ either static or dynamic *channel allocation* to associate nodes with channels. The channel allocation schemes used by MAC protocols can be categorized into two basic types: contention based (we use the term contention or random access interchangeably) or contention-free (no possibility of packet collisions).

The aim of this chapter is to provide insights into key aspects of the MAC problem and options available to the designer of a UAN. We revisit some of the results from the terrestrial wireless domain to evaluate its applicability to UANs. It is important to have a holistic picture of the various aspects of the problem so that the designer can make an informed choice based on the exact requirements of the network being set up. It is mentioned in many UAN papers that FDMA is inefficient for UANs.[3] Static TDMA is quite widely used in many UANs,[4] and CDMA is often highly recommended over TDMA and FDMA.[5] We take a look at these choices and evaluate their relative merits.

The rest of this chapter is organized as follows. The remaining part of this section establishes some basic concepts and terminology. In Section 8.2, we look at the different channelization options of time, frequency, and code division and illustrate why time division is a good choice for UANs. MACA[6]-based protocols are then illustrated as a special case of dynamic TDMA, and are shown to address the UAN MAC problem well. We then look at some of the other important aspects of MACA-based protocols in UANs. We present a simple performance analysis for MACA-based protocols in Section 8.0.

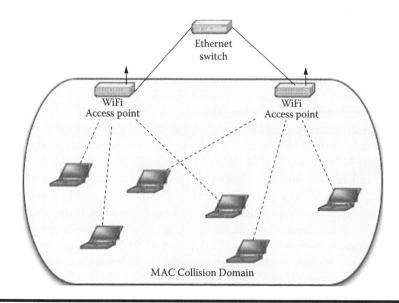

Figure 8.1 **Two overlapping WLAN networks connected via an Ethernet switch illustrate the difference between a MAC topology and a routing topology. The MAC topology in this example is distributed, while the routing topology is a tree.**

8.2 Topology

The logical topology of a network adopted by a MAC protocol can potentially be different from that adopted by the higher layers for routing. For example, consider two typical wireless local area network (WLAN) access points operating on the same channel in an overlapping geographical area. If the WLAN access points are connected via an Ethernet switch as shown in Figure 8.1, they can route packets between their respective clients. In terms of routing, this is a tree topology. However, there is no central control of medium access, and therefore the MAC topology is distributed. In this chapter, we are interested in the MAC topology and not the routing topology.

In a *distributed topology* (sometimes referred to as peer-to-peer), there are no controlling central nodes and all nodes asynchronously and equally handle MAC functionality. All nodes are deemed equal in the MAC function. The IEEE 802.11 distributed coordination function (DCF) in ad hoc mode is a distributed protocol used by WLAN networks.[7] The 802.11 DCF protocol can be used for peer-to-peer networks in ad hoc mode.

In a *centralized topology* (also referred to as clustered, cellular, etc.), the central node controls medium access for nodes in its neighborhood and is tasked with allocation of data channels to client nodes. For example, in a cellular wireless network, only the base station (BS) has the MAC and routing gateway functionalities.

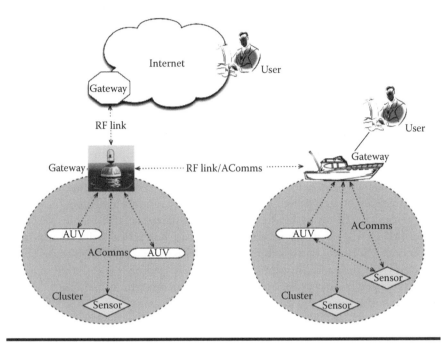

Figure 8.2 An example UAN architecture with two cells, one using a central-ized topology while the other using a distributed topology. (From M. Chitre, S. Shahabudeen, and M. Stojanovic, *Marine Technology Society Journal—"The State of Technology in 2008,"* vol. 42, Spring 2008. With permission.)

In such a network, ordinary nodes only communicate with such BSs. The 802.11 point coordination function (PCF) is a centralized topology protocol and the access point (AP) polls and allocates channels to clients.[7] The 802.11 PCF is implemented on top of DCF. The AP acquires channels in a distributed fashion among its peers and utilizes the acquired channels for nodes under its PCF control. So it can be viewed as a centralized topology within a distributed topology!

We define a *cell* as an area under the control of one central controller in a cen-tralized topology. In a distributed topology there are no defined cells, but as nodes get farther apart they will not interfere with each other due to signal attenuation.

Figure 8.2 shows a possible UAN. This network consists of two cells, one using a centralized topology (all communications are between sub-sea nodes and a cen-tral surface BS) and the other using a distributed topology (there are links between sub-sea nodes directly). The UAN cells in this example are interconnected via sur-face radio and to the wider Internet. Centralized topology is a good choice when at least one node in the cell has high-speed connectivity to other cells. However, in many UANs, all the nodes have similar communication capabilities. The distrib-uted topology is thus more applicable to such scenarios and is the main focus of this chapter. But we also discuss centralized topology aspects alongside.

8.2.1 Spatial Re-Use, Channelization, and Allocation

In a centralized topology, the total available channel capacity has to be divided spatially between cells for re-use purposes. If the network spans a geographic area larger than single hop range of a single node, channels can and need to be re-used due to limited channel capacity. This is commonly done in cellular networks. Re-use patterns are designed to allow for maximum permissible co-channel interference. Re-use patterns and inter-cell channelization can be achieved dynamically or statically. In 802.11, 11 fixed frequency bands are allocated for use in cells (only three are actually orthogonal) and a given AP uses a fixed frequency band. Thus there is a static frequency band-based channelization between cells. In certain cellular networks, there are provisions for dynamic inter-cell channelization through concepts such as channel borrowing between base stations supervised by a mobile switching center (MSC).[8] In a distributed topology, there are no explicit "cells," but the network has to effectively do some form of channelization and spatial re-use of channels over large areas and this is further discussed in Section 8.4.

Within a cell, the channel capacity needs to be divided for multiple access. In many networks, the channelization within a cell is essentially static. Cellular wireless systems such as GSM (Global System for Mobile Communications) allocate pre-defined TDMA time slots and frequency bands to users upon request and hence can be considered static channelization within a cell. In cellular systems, multiple channels also can be allocated to the same user, for example, HSCSD (High-Speed Circuit-Switched Data) mode in GSM and can be considered a form of dynamic channelization within a cell. Note that the static channels created by the channelization are dynamically allocated to different users, so the channel allocation is dynamic. In distributed topology protocols such as 802.11 DCF, the length of data can be varied. This in effect varies the allocated time slot and can be considered as a form of dynamic TDMA channelization within a cell.

In static channel allocation protocols, nodes are permanently allocated predetermined channels (also referred to as scheduled or deterministic protocols). In dynamic channel allocation protocols, one option is to use pure random access (such as ALOHA), i.e., with no explicit channel reservation. Another option is to use random access only for the control channel; the *control channel* is a shared channel that all nodes can send and listen to, and use to make reservations for *data channels*. For example, in dynamic distributed topologies, nodes can request for a data channel using a random access protocol (on the control channel) and are assigned a data channel by the recipient. As another example, in dynamic centralized topologies, nodes can send requests for a data channel to the central controller using a random access control channel. In dynamic centralized networks, a third option is to have no contention even in the control channel, for example, the central controller can poll the clients over a control channel. As discussed above, the allocated data channel can be statically channelized or dynamically channelized (as in HSCSD in GSM or 802.11 DCF).

Also, general UANs require bi-directional communications. In a distributed topology, there is bi-directional symmetry and no additional considerations are required. In centralized topology, we can have duplexing in time, code, or frequency. However, most existing UAN physical layers are half duplex due to practical considerations. This has a significant impact on design choices in UANs and is discussed further in Section 8.4.

8.2.2 General Protocol Models for Dynamic Data Channel Allocation

In a distributed topology each user has to inform the recipient as well as neighbors about its intention to transmit data. In one of the most common paradigms, a node sends out a request to send (RTS) over the control channel. The receiver responds with a clear to send (CTS) over the control channel indicating the allocated data channel. Neighbors note the channel allocation. In a distributed topology, we cannot in general avoid contention for the control channel. In such a contention-based RTS-CTS scheme, collisions are possible. The standard techniques used to reduce collisions include random back-off before sending RTS using either a constant window or a freezing back-off.[7] When collisions occur, RTS or CTS packets are lost and the transmitter resends the RTS after random back-off.

In a centralized topology, a contention-based shared control channel can be used to communicate with the central node using similar RTS-CTS exchanges as above. This is the basic model in GSM cellular networks even though the terminology used is different. A UAN that employed this model was an acoustic local area network (ALAN) deployed in Monterey Canyon to perform long-term data acquisition and ocean monitoring from multiple bottom-mounted nodes.[9,10] The subsurface nodes send asynchronous requests via a shared channel (similar to RTS) to send data packets to the master node. The master node sends an acknowledgment (similar to CTS) via a different channel indicating the data channel to be used. The node transmits data immediately on the data channel. The channelization for request, acknowledgments, and data is in the frequency domain. As is the case for 802.11 DCF in ad hoc (distributed topology) and infrastructure (centralized topology) modes, the same basic MAC protocol can be used in both topologies.

Contention-free polling is a good option for a centralized MAC topology. The central node polls ordinary nodes and allocates a data channel. The 802.11 PCF is a good example. In another UAN example,[11] the polling-based protocol called FAMA-CF (Floor Acquisition Multiple-Access–Collision Free) uses request for RTS (RRTS), RTS, CTS, DATA, ACK handshaking to communicate with the central node. The central node initiates the RRTS to its peers.

Pure random access schemes for data transfer (such as ALOHA) waste channel capacity due to excessive collisions. The maximum theoretical throughput of classical ALOHA is only 18%. However, variations of such random access protocols can be considered for short-duration data transfer and are discussed in Section 8.4. For example, 802.11 DCF defines a "basic access scheme" that does not use RTS-CTS, but instead uses a contention-based DATA-ACK protocol.

Special Note: Variations of ALOHA could also be a good choice for underwater network deployments with large latency (node separations of tens of kilometers or more as a single collision domain) where the handshaking of RTS/CTS protocols can take up significant time-bandwidth. But most acoustic modems (with a usable data rate for networking) employed in underwater networks today have only a few kilometers of range, thereby limiting a single collision domain to a few kilometers. Thus with latencies ranging from less than one second to a maximum of a few seconds, RTS/CTS-based mechanisms are a very good choice in many practical underwater networks. This is further discussed in Section 8.4.

8.2.3 Need for Dynamic Channelization and Allocation in UANs

Channelization of any kind implies division of some allowed maximum channel capacity. Maximum usable frequency bandwidth and hence the channel capacity could be restricted by regulation and/or other physical layer requirements. In inter-cell channelization, if base stations use only one fixed channel as in 802.11, then it implies usage of only a portion of available capacity by a certain cell. If there are no neighboring cells, such a situation is not good for capacity-starved UANs. Therefore, inter-cell channelization needs to be dynamic.

In simple UANs, where a fixed number of nodes are deployed and there are sufficient channels, one can use static protocols regardless of utilization level in each channel. Static TDMA, FDMA, or CDMA protocols can easily be used for small static networks based on the capabilities of the underlying physical layer.

In the case of a large number of nodes, a channel per node requirement could yield very low data rate channels. If each node only transmits occasionally, the total capacity could be used more efficiently by having a smaller number of high data rate channels. If there are more nodes than there are channels, once all the channels are allocated, further requests are blocked. We need to be able to provide channels to all users without blocking. We therefore need protocols to dynamically channelize the intra-cell channel capacity also and allocate channels on demand. Capacity of a single channel needs to be adaptively varied according to number of nodes. Channels should be allocated only for short periods of time (this is a form of channelization in time) and nodes need to contend for them repeatedly.

Another reason to consider dynamic allocation in MAC protocols is the need to allocate channels on demand in response to arrivals and departures of nodes. This is the case for cellular mobile networks with client nodes wanting to make calls. In UANs, such cases could arise in AUV networks.

8.3 Selection of MAC Protocols for UANs

Based on the concepts discussed in the previous section, we outline the selection of an effective MAC protocol for use in UANs. Dynamic channelization, channel allocation, and the selection of appropriate topologies are all important issues to be addressed. But first we need to select appropriate channelization—code, frequency, or time division. There has been much discussion about the relative merits of CDMA, FDMA, TDMA, and protocols such as MACA.[5] We discuss these channelization options to illustrate their relative merits.

8.3.1 The General Equivalence of Static TDMA, FDMA, and CDMA

Static FDMA, TDMA, and CDMA effectively provide same unidirectional data rate performance in an ideal case. To make comparisons on equal terms, let us assume that the system has a fixed maximum bandwidth B. The average power consumption over time must be the same for all systems for fair comparison, and to that effect we assume maximum transmission power P_T per unit bandwidth. Let P_R be the received power per unit bandwidth. We assume Gaussian ambient noise with power N_o per unit bandwidth. We consider the data transmitted in a time window of length T. Let N be the number of user channels and R be the data rate. Let γ be the received signal-to-noise ratio. The received signal-to-noise ratio over the bandwidth B is the ratio of the received power $P_R B$ and the ambient noise power $N_o B$. For simplicity, we assume that all links experience equal frequency-independent path-loss ratio β and hence $P_R = \beta\, P_T$ remains a constant.

The theoretical capacity for a given additive white Gaussian noise (AWGN) communication channel is given by Shannon's law as

$$R = B \log_2 (1 + \gamma) \tag{8.1}$$

We use the law to illustrate the equivalence of R from an information theoretic standpoint. First let's compare TDMA with FDMA. For a TDMA system, each user time slot is T/N time units long and uses the entire bandwidth B at maximum power P_T. Therefore the effective data rate per user over time T is

$$R_{TDMA} = B \log_2 \left(1 + \frac{P_R B}{N_o B} \right) \frac{T}{N} \frac{1}{T}$$

$$= \frac{B}{N} \log_2 \left(1 + \frac{P_R}{N_o} \right) \tag{8.2}$$

For an FDMA system, each user time slot is T time units long and uses B/N units of bandwidth at maximum power P_T. The effective data rate over time T is

$$
\begin{aligned}
R_{FDMA} &= \frac{B}{N} \log_2 \left(1 + \frac{P_R(B/N)}{N_o(B/N)} \right) \\
&= \frac{B}{N} \log_2 \left(1 + \frac{P_R}{N_o} \right)
\end{aligned}
\tag{8.3}
$$

This is the same as R_{TDMA} and is the same equivalence embodied in the familiar concept of fixed time-bandwidth product. This implies that in a fixed amount of time T, N users using N bands of B/N get the same average data rate as N users dividing T into N slots of T/N and each using the entire bandwidth B.

Let us now look at ideal CDMA utilizing a total bandwidth B after spreading. The spreading factor k expands the bandwidth by k. To be able to compare on equal terms, we use a fully loaded orthogonal CDMA system with $k = N$. The bandwidth per user is B/k units before spreading. Considering orthogonality of CDMA codes, i.e., multiple access interference (MAI) is zero (requires perfect time synchronization), Shannon's law for the de-spread channel can be re-written as

$$
\begin{aligned}
R_{CDMA} &= \frac{B}{k} \log_2 \left(1 + \frac{P_R(B/k)}{N_o(B/k) + MAI} \right) \\
&= \frac{B}{k} \log_2 \left(1 + \frac{P_R}{N_o} \right)
\end{aligned}
\tag{8.4}
$$

If $k = N$ for a fully loaded synchronous orthogonal CDMA,

$$
R_{CDMA} = \frac{B}{N} \log_2 \left(1 + \frac{P_R}{N_o} \right)
\tag{8.5}
$$

Thus this is exactly the same as the FDMA and TDMA ideal cases. Such equivalences have been discussed in prior publications.[12]

It is interesting to note that channelization using CDMA or FDMA also requires division in time as all transmission frames are finite in time. In other words, we can consider TDMA as a fundamental division mechanism for all channels. FDMA and CDMA could be viewed as further orthogonal divisions in frequency or code space.

8.3.2 General Strengths and Weakness of CDMA, FDMA, and TDMA

The above comparison between TDMA, FDMA, and CDMA are true under some ideal assumptions. TDMA, for example, has inefficiencies arising due to required guard periods between slots to compensate for imperfect clock synchronization between nodes, clock drift, and latency. FDMA requires guard bands between bands, since infinitely sharp cut-off filters are physically impossible to realize.

In CDMA, the above result holds only when all chips are synchronized and mutually orthogonal. It works well in downlink synchronous transmissions in terrestrial radio wireless networks. In high latency UANs, however, this is not easy to achieve. There will be interference between channels due to non-orthogonality of codes arising mainly from latencies. Also, typical CDMA analysis assumes perfect power control. The classical near-far problem will otherwise create large MAI from near-by sources and performance will degrade.

Considering MAI from M other users (based on published analysis[8]) for asynchronous CDMA, Shannon's law for the de-spread channel can be re-written as

$$R_{CDMA} = \frac{B}{k} \log_2 \left(1 + \frac{P_R(B/k)}{N_o(B/k) + M(P_R/k)(B/k)} \right)$$
$$= \frac{B}{k} \log_2 \left(1 + \frac{P_R}{N_o + M(P_R/k)} \right) \tag{8.6}$$

If $k = N$ for a fully loaded CDMA,

$$R_{CDMA} = \frac{B}{N} \log_2 \left(1 + \frac{P_R}{N_o + M(P_R/N)} \right) \tag{8.7}$$

M can be at maximum $N-1$ in an N-node fully connected scenario. As M increases, MAI increases and performance degrades to less than the equivalent FDMA or TDMA protocol. Such degradation has been shown to make asynchronous CDMA (as would be the case in distributed topology underwater networks) less attractive than FDMA or TDMA in a fully connected system of N nodes.[12]

An advantage of TDMA-based protocols is that they provide flexibility in terms of implementation over any Physical Layer technology. As long as the MAC layer has access to transmit and receive behaviors of the Physical Layer, any underlying physical layer such as OFDM (Orthogonal Frequency-Division Multiplexing), FH-BFSK (Frequency-Hop/Binary Phase-Shift Keying), etc., may be used.

8.3.3 Full Duplex Requirement for CDMA and FDMA

In the case of cellular wireless networks using static FDMA or CDMA, the clients communicate with a multi-band or multi-code base station. Thus all users can communicate in parallel with the base station, and vice versa. The general equivalence in Section 8.4 is thus valid.

Let us consider the case of N isolated nodes communicating with each other in a distributed topology. In TDMA, since only one node is active at a given time, it can transmit to any one or more of the $N-1$ neighbors. For FDMA or CDMA, we need to assume that they are all transmitting simultaneously in orthogonal frequency or code bands to get the same or similar performance as the TDMA. So unless all nodes can receive on all the channels (frequency bands or CDMA codes) while transmitting, there can be no receivers! This implies that all nodes are capable of receiving all bands or codes (except the one in which it transmits) in parallel and at the same time transmit, i.e., each node is full duplex.

Thus, the key difference with TDMA is that nodes need to be able to receive while they are transmitting (full duplex). In CDMA this could amount to a near-far problem and reduce the performance. In FDMA extremely good inter-band filtering would be required to minimize bandwidth wastage due to guard bands. Also, typically packets use a detection preamble followed by data.[13] So unless the preamble is also code or frequency band tunable, there will be interference from the preamble in code division and frequency division data channel methods. Smaller bandwidth for preambles reduces detection performance. Thus such multi-band full duplex receivers are typically not used and most underwater modems available today are half-duplex. There has nevertheless been some work on full duplex systems in UANs.[14]

8.3.4 Additional Discussion on CDMA

As discussed earlier, CDMA can be shown to be inferior when used in a single cell or among N fully connected nodes.[12] However, traditional analysis shows that CDMA has performance gains in cellular networks with re-use considerations. In traditional TDMA- or FDMA-based systems, a re-use pattern of seven is commonly used. In CDMA, frequency re-use is not necessary (although sometimes used), and when using all CDMA codes in a single cell and considering co-channel interference from neighboring cells, capacity gains over TDMA or FDMA of up to five times have been shown.[12]

However, we will take a closer look at some of this analysis to see if comparisons were made on equal terms and are appropriate for UANs. In analysis such as in,[12] a voice activity factor of about 3/8 is used in asynchronous CDMA capacity analysis, but is not applied to GSM or TDMA analysis. With B as the total bandwidth, and T as time window as defined in Section 8.4 (BT is the time-bandwidth product), and if K_{CDMA} is the number of channels available (and used) in a cell (for CDMA), using the analysis without the voice activity factor, the signal-to-interference-noise ratio (SINR) ψ can be written as[12]

$$\Psi = \frac{2BT}{1.5K_{CDMA} - 1} \tag{8.8}$$

The above considers interference from six neighboring cells. This gives an upper bound to the number of channels for a given SINR as

$$K_{CDMA} \leq \frac{4BT}{3\Psi} + \frac{2}{3} \tag{8.9}$$

If ψ is 5 (i.e., 7 dB)[12],

$$K_{CDMA} \leq \frac{4BT}{15} + \frac{2}{3} \tag{8.10}$$

A comparison for an FDMA-based system[12] uses a re-use factor of seven, and the number of channels in a cell K_{GSM} is

$$K_{GSM} = \frac{BT}{7} \tag{8.11}$$

Thus, with the voice activity factor discounted, K_{CDMA} is shown to be about 1.75 times K_{GSM} (assuming $BT \gg 2/3$). It is true that CDMA has an advantage in being able to use voice activity factor for capacity increase in voice traffic, as this cannot directly be done in FDMA/TDMA-based systems. We note that for data communications, the benefit of such a factor will depend on the burstiness of traffic.

However, in standard hexagonal-based geometry analysis[8] used for FDMA- or TDMA-based systems, a cluster size of seven gives an SINR of 18 dB! Using 18 dB ($\psi = 63$) as the criterion in the CDMA Equation (8.9),

$$K_{CDMA} \leq \frac{4BT}{180} + \frac{2}{3} \tag{8.12}$$

This shows that under similar SINR requirements, the TDMA or FDMA system has much better performance than asynchronous CDMA. Also, perfect power control is assumed in such typical CDMA analysis[12] and CDMA performance significantly varies with power control errors.[15]

For fourth generation (4G) OFDM-based systems, TDMA has been considered a better multiple access option compared to FDMA and CDMA.[16] WiMAX, a popular new technology, uses orthogonal frequency division multiple access (OFDMA), where OFDM sub-carriers are assigned as channels to users, and can be interpreted as a form of dynamic FDMA. OFDM has been shown to be better than CDMA in multi-path environments[17] and now seems to be the choice for 4G communications.[16] UAN modems using OFDM technology have also been developed.[18]

The above discussion is primarily to say that even in terrestrial cellular systems, CDMA's superiority is arguable. Each system has its own difficulties. CDMA eases re-use planning, but there are real world issues such as power control and code offsets between cells to separate users' pseudo-random number (PN) sequences.[15] Moreover, the primary focus of this chapter is on distributed topology and the full-duplex and multi-channel requirements in CDMA and FDMA make TDMA a better choice of channelization for distributed topology UANs.

8.4 Dynamic Allocation Protocols

Having established that TDMA fundamentally is a good choice for UANs in distributed topology, we next look at a dynamic variant (henceforth referred to as dynamic TDMA or D-TDMA) in order to address the dynamic channelization and allocation requirements that we set out at the beginning. This can be modeled using a frame with multiple slots, with a contention period between frames as shown in Figure 8.3(a). Nodes use random access during the contention slots (control channel) to request data slot (data channel) allocations in the frame. A channel is allocated for a certain number of frames (a system parameter). Once all the channels are allocated, further requests are not satisfied until channels are relinquished. Multiple successful contentions can happen within a single contention slot. Similar models for dynamic TDMA are used in many terrestrial networks such as Hiperlan/2, which was a parallel development to 802.11 standards by the European Telecommunications Standards Institute (ETSI).[19]

For comparison, dynamic FDMA or CDMA equivalents are shown in Figure 8.3(b). The control channel for contention can be a band or code by itself (although the requests in a frame may be for allocation of channel in subsequent frames). The same equivalence as described in Section 8.3 holds except for the additional shared contention-based control channel. Just like the contention frequency band in FDMA or control channel code in CDMA takes up available channel capacity, the contention period in the dynamic TDMA takes up a certain portion of total channel capacity. In terms of capacity, this control channel is equivalent irrespective of the type of channelization used.

Any of the above protocols can easily be implemented in centralized networks (they are used in cellular wireless networks). In distributed topology networks, if a node receives an RTS, it can allocate one or more of the unused receive channels to the requestor. Nodes need to be aware of what channels are being used in the

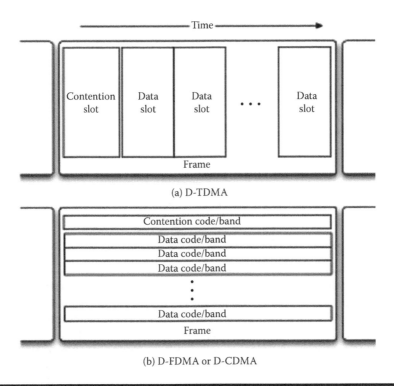

Figure 8.3 A schematic representation of D-TDMA, D-FDMA, and D-CDMA.

neighborhood to avoid collisions. Note that all receivers have to be full duplex multi-band or multi-code capable for this to work equally well in distributed FDMA or CDMA systems. As discussed earlier in Section 8.3, this is not typically preferred for UAN systems. The D-TDMA protocol works well without such requirements.

D-TDMA has a contention slot unlike pure TDMA that takes up bandwidth and reduces efficiency. This is the primary cost for the ad hoc capability that D-TDMA offers.

8.4.1 Dynamic TDMA Protocol and MACA-Based Protocols

We wish to show how the MACA protocol relates to the D-TDMA protocol described above. To begin, let us look at the key details of MACA-based protocols. When we use the term MACA, we are referring to a family of closely associated protocols that uses essentially the same principles of handshaking, etc., as MACA. MACA uses the general model described in Section 8.3. The transmitter sends an RTS to the receiver and the receiver responds with a CTS. Upon reception of the CTS, the transmitter sends the DATA packets in a batch (referred to as DATA_TRAIN) and the number of packets in a batch is variable and is specified in RTS.

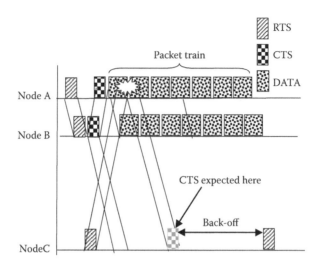

Figure 8.4 MACA protocol model with RTS/CTS/PACKET_TRAIN illustrating some of the key aspects. Node A sends an RTS to Node B and Node B sends a CTS back to Node A. Node A then sends a DATA batch to Node B. Reception of CTS at another node C is shown, which then performs a VCS to avoid interference with Node A's transmission. A potential collision from Node C is shown. How back-off starts after completion of one batch transmission is also indicated. (From S. Shahabudeen, M. Chitre, and M. Motani, "A multi-channel MAC protocol for AUV networks," in *IEEE Oceans' 07*, Aberdeen, Scotland, 2007. With permission.)

If CTS is not received, the transmitter does a random back-off and repeats the process. Once the receiver successfully receives some part of the data train, it sends an optional acknowledgment (ACK). Figure 8.4 illustrates this. More protocol details and performance analysis for MACA are given in Section 8.3.

Next, we relate MACA to the D-TDMA protocol. In D-TDMA, let all the slots in a frame be allocated to one successful requestor at a time and let there be a variable number of slots in a frame. Let each slot be viewed as a DATA packet and a frame as a DATA_TRAIN as used in the MACA-based protocol. Let the contention slot be also of variable duration as determined by the completion of one successful RTS/CTS exchange. This is the same model as MACA discussed above! Thus MACA can be viewed as a special case of a D-TDMA scheme and inherits many of the stated advantages. The most important advantage is the requirement of only a half-duplex physical layer. D-TDMA is very similar to many of the efficient dynamic protocols used in cellular networks. These protocols employ random access for the control channel only. Usually MACA is viewed as a different class of protocol, a type of random access protocol. Through the above illustration, we have shown that D-TDMA and MACA are closely related, and that the two apparent classes of protocols are not so different after all. In fact, it's interesting to note that the almost parallel development of Hiperlan/2 protocol by ETSI and 802.11

by IEEE used D-TDMA and MACA-based protocols, respectively.[19] Of the two, IEEE 802.11 is the more commonly used wireless LAN protocol today.

By allowing variable data length, MACA-based protocols can allow a variable number of nodes to communicate within the same time frame and thus effectively meet the requirement to have dynamic channelization. There is no strict upper limit on the number of users it can support as the duration of the data frame can be varied. Performance gracefully degrades with number of users. MACA also eliminates the critical difficulty with clock synchronization required by D-TDMA as there are no repeated frames with multiple time slots for nodes to deal with. Other technical difficulties with FDMA and CDMA outlined in Section 8.3 are also absent. Thus MACA provides the basis for one of the most flexible, robust protocols for a dynamic channelization and allocation that works well in distributed topology and can be extended to centralized topologies as in 802.11 PCF.

MACA-based protocols were shown to be effective for underwater use early on in the Seaweb project.[3] The authors observe that in the physical and MAC layers, adaptive modulation and power control are critical to maximize both channel capacity and efficiency and RTS/CTS handshaking permits that, along with other benefits such as addressing, ranging, etc. Power-minimized nearest-neighbor routing has been shown to be the best option in a distributed scenario for a variant of MACA.[20] The optimal power is found to be that which minimizes connectivity. MACA-based protocols are found to be highly suited in many scenarios in UANs where scalability is important and time-synchronization is not available.[1,11,21,22]

We next look at the benefit from the use of packet trains. The classic hidden node collision problem of MACA[21] is shown in Figure 8.4. If instead of the packet train a single large DATA packet is used, in such an RTS collision, the entire DATA packet is more easily lost. And for re-transmission, the complete RTS, CTS, DATA, ACK exchange has to be repeated, needlessly wasting channel capacity. Also, arbitrarily long-duration single-coded packets might not be feasible due to Physical Layer memory and processing limitations. When packet trains are used, the RTS collisions only affect some of the packets in the train and the ACK will indicate this. By using a fairly large number of packets in the train, throughput efficiency can be greatly improved as the results demonstrate in Section 8.5. The idea that packet trains improve performance of protocols such as MACA can be found in other papers.[21,23]

8.4.2 Re-Use, Topology Selection, and 802.11 Applicability

In a distributed topology, when using MACA-based protocols (as used in 802.11 DCF ad hoc mode), when links (two nodes communicating with each other) are separated sufficiently in space, we can infer that some links can operate simultaneously in time. Thus, in other words, time domain channel re-use is inherent. Re-use happens automatically and no extra mechanisms are needed. Thus for UANs where contiguous multi-hop nodes are present, we advocate the use of distributed protocols along the principles of 802.11 DCF, and not the use of pure centralized topology (note that the centralized 802.11 PCF operates within the DCF framework). Centralized topology is best only for situations operating as a single collision domain or in the case of spatially separate collision domains connected via surface radio gateways. This is the case in terrestrial cellular wireless networks where the BS are connected to each other through a high-speed wired network. If centralized topology protocols follow the approach of 802.11 PCF, since it essentially rides on DCF, the same automatic re-use mechanisms are present and can also be used.

In essence, 802.11 WLAN concepts are quite suited for use in UANs. DCF is the most popular option in terrestrial network implementations; PCF is not very popular and is not supported by many WLAN devices. However, there are significant differences in latencies in a UAN as compared to a WLAN. This has a significant impact on protocol features such as physical carrier sense (PCS) as used in 802.11. Also in channel-capacity-starved UANs, inter-cell channelization used in 802.11 (802.11 has 11 frequency bands for use in different cells) would not be advisable.

8.4.3 Other Aspects of MACA-Based Protocols

In this section, we first discuss the long latencies common in UANs—this is a commonly cited issue for MACA-based protocols operating under water. Following that we look at some of the extensions to MACA protocols that have been proposed in the UAN literature and some of the modes and service that the data-link layer (DLL) needs to offer to the higher layers. There are some known issues such as starvation in MACA-based networks[24] that we do not cover in this chapter. Starvation modeling and prevention remains an open research area in such networks.

We note that the MACA contention period is variable in duration, unlike D-TDMA, and hence reduces efficiency and introduces variable data transfer delay. These are costs for the increased flexibility and robustness that MACA offers (not requiring precise time synchronization).

8.4.3.1 Latency and Its Impact

How does latency affect the behavior of a MACA-based protocol? The basic impact of latency is the loss of channel utilization and efficiency in the transmission delays between RTS, CTS, DATA, and ACK packets. A special note in Section 8.3 briefly discussed the issue of latency in underwater networks. From the easily established fact that most commercial acoustic modems with a useful data rate have only a few kilometers of range, we can infer that most practical underwater acoustic networks have a few kilometers as a single collision domain at maximum. Thus, as mentioned in Section 8.3, RTS/CTS-based protocols are not heavily impaired in most cases. As we shall see in Section 8.3, the use of appropriate batch sizes in data packet train can help counter this loss of efficiency.

Also, such loss of efficiency is inherent in the dynamic channel allocation model described in Section 8.3 and is a fundamental limitation of dynamic ad hoc networks that satisfies the requirement of dynamic channelization and allocation as outlined at the beginning of the chapter. Only static networks can avoid this loss of efficiency. Thus, it is not a problem with MACA per se, but a limitation arising directly from network requirements and underwater characteristics.

However, as noted in Section 8.3, in the case of very high latency networks that span tens of kilometers as a single collision domain, ALOHA-based variants with no handshaking could be explored, but is not in the scope of this chapter. Also, there are some variations to MACA-based protocols being proposed[25] to counter the high latency. However, actual field trials of such proposals have not been reported.

MACA in its original form does not have PCS. However, if PCS were used as in 802.11, its effectiveness would be undermined by high underwater latency. PCS works on the premise that when a node transmits, all the other nodes hear it instantaneously. In UANs this is not true. The authors' own simulation studies have shown that carrier sensing makes only negligible differences in performance. There are published variants of MACA that use PCS (e.g., Slotted-FAMA[21]). However, the efficacy of PCS in UANs has not been conclusively shown. Some of these are still open research problems.

8.4.3.2 Protocol Extensions and Enhancements of MACA

Many protocol extensions and enhancements to MACA have been investigated to improve the suitability of MACA to UANs. For example, a WAIT command extension has been investigated.[4,26] A WAIT command is sent back by the receiver if it is currently busy and intends to send CTS later on. In another concept,[3] instead of using ACK packets, selective ARQ (Automatic Repeat-Request) can be initiated by the recipient should it not receive packets in a specified time. Distance aware-collision avoidance protocol (DACAP) is also based on MACA.[27] It adds a warning message if an RTS is overheard while waiting for a reply to its own RTS. While waiting for a reply, if another CTS or a warning is heard, a random back-off is used. Floor acquisition multiple access (FAMA), a family of protocols of which MACA is a variant, was originally proposed for terrestrial networks. FAMA uses PCS (absent in MACA) and puts restrictions on RTS/CTS time durations. FAMA in its original form is quite unsuited to UANs, but with suitable modifications, it can be effectively used under water.[21] Such enhancements of MACA-based protocols for UANs are a subject of ongoing research as shown in a recent survey.[28]

We also note that there are some protocol proposals for underwater networks such as T-LOHI, which aims at collision avoidance, but is not directly derived from MACA concepts.[28,29] Such protocols hold promise but no actual deployments have been reported. Their efficacy in real sea environments with channel characteristics such as reverberations, etc., are yet to be proven. In contrast, MACA-based schemes have been successfully demonstrated in actual sea trials using acoustic modems for nearly a decade.[30,31]

8.4.3.3 Different DLL Modes and Services

The DLL, which includes the critical sub-function of MAC, also needs to cater for different message classes such as reliable, unreliable, broadcast, and unicast messaging. Reliable messaging could be defined as messaging where re-transmissions upon failure will be handled by the DLL. For unreliable messaging, there are no re-transmissions by the DLL. Both these types of messaging fit into the RTS, CTS, DATA_TRAIN, ACK model discussed earlier.

DLL can also support messages without RTS/CTS handshaking. DATA packets can be sent using the same back-off logic as RTS and this is very closely related to the ALOHA model. This is the "basic access scheme" in 802.11 DCF. For low load situations, this can yield lower round-trip latencies by avoiding the handshaking, and as mentioned earlier could also be suited for very large latency scenarios (Section 8.3). In other words MACA-based protocols can be operated in different modes to provide different service classes to the higher layer.

8.5 Performance Analysis

In this section we present a simplified file transfer performance analysis of the MACA-based protocol in a distributed topology with low to medium latency (see notes in Section 8.4). In many sensor networks, the collected data has to be transferred as files across an acoustic network. Hence we analyze the performance of a file transfer operation. Many other types of queuing analysis are possible for arrival traffic with distributions such as Poisson. The presentation here is aimed at giving the reader a flavor for analysis techniques and a feel for MACA-based protocols. The pure centralized topology is also beyond the scope of this chapter (some performance analyses for centralized topology UAN protocols have been published[11]). The analysis below helps capture the essentials of MACA behavior for quick understanding of expected performance in a single-hop network. Interested readers can find alternative analysis by the authors[31] and other analysis for similar protocols used in underwater networks.[22]

In some of the papers on analysis of similar protocols in terrestrial radio networks,[7] the following performance measure is termed as "saturation throughput"—the throughput of the network when the queue is saturated or always has data to transmit. Such a measure is valid for file transfer applications. We define normalized throughput T as the number of packets successfully transferred per unit time normalized by the system capacity. Let the packet length be L and the batch size be B. Hence the system capacity is one packet in time L. Other parameters include number of neighbor nodes N that each node is effectively in contention with, maximum propagation delay D, back-off window size W, and back-off and retry mechanism (we use uniform back-off and an infinite retry model). Let t represent time. Timeout t_A is used to wait for CTS and ACK. Timeout t_A is related to D and to control packet length L to give enough time for the round trip delay as

$$t_A = 2D + L \qquad (8.13)$$

If P_d is the packet detection probability and P is the probability of correctly decoding a packet, the probability k that a packet is detected and decoded correctly is

$$k = P_d P \qquad (8.14)$$

In the following analysis, we do not consider hidden nodes for simplicity.

8.5.1 Algorithm Outline

Though not necessary for the MACA protocol presented in this chapter, it is generally known that time slotting improves the performance of contention-based protocols (e.g., slotted-FAMA[21]). Here we assume time slotting primarily to simplify analysis.

During RTS contention phase, the slot duration l is defined as

$$l = L + D \qquad (8.15)$$

This allows for collisions to be contained within the slot boundaries. As noted before, this model is for low latency underwater networks and it works best for D less than 1 second or so (roughly 1.5 km maximum network span) and packet size L is of the order of 0.5 seconds. In the RTS contention algorithm used here, a node starts with uniformly selected back-off time slots in the integer range of $[1, W]$. The actual contention window time period is Wl. When the back-off timer expires, an RTS is sent. Once RTS is sent, CTS timer t_A starts. If the timer expires before reception of CTS, RTS back-off procedure starts again. Once CTS is received, DATA train is sent followed by wait for ACK. If ACK is not received, the RTS cycle repeats. Reception of RTS/CTS packets and a possible DATA frame while waiting to send RTS triggers Virtual Carrier Sense (VCS). Successful DATA transmission for any one node restarts RTS contention cycle for all. Note that 802.11 uses freezing back-off, which is described in [7]. This protocol does not use PCS.

8.5.2 Expected Time to Successful RTS

Firstly, we find W_A, the expected delay from start of RTS contention until an RTS is sent from a given node with no collision. W_A does not include time until correct reception of the RTS at the receiver or the reception of CTS at the transmitter (in terms of detection and decoding, i.e., W_A deals only with collisions).

Let τ be the probability that at least one node transmits an RTS in a given slot and it encounters no collision from any other node (i.e., probability of collision-free RTS transmission in a given slot). All nodes use the same contention window W at any given time.

Assuming independence of RTS transmission among nodes

$$\tau = N \frac{1}{W}\left(1 - \frac{1}{W}\right)^{N-1}$$
(8.16)

If we assume that the probability τ in any one slot is independent of the previous one, we can model τ as a Geometric random variable. W_x, the expected time until any one node transmits an RTS without collision, is

$$W_x = \frac{l}{\tau}$$
(8.17)

In the following analysis, we look at the problem from the perspective of a given node, which we shall refer to as the "current node." As shown in Figure 8.5, when any one node transmits an RTS, let P_Y be the probability that it is the current node. The current node then waits for CTS. As all nodes are equally likely to succeed,

$$P_Y = 1/N$$
(8.18)

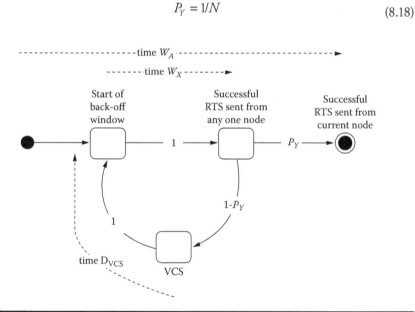

Figure 8.5 **Model to compute expected time to successful RTS.**

The probability that it is not the current node is $1-P_Y$. The probability that the current node receives that RTS is k, in which case it goes into VCS. It will wait for CTS for a minimum of t_A time units. Let P_{CTS} be defined as the probability of getting CTS for this received RTS. Since the recipient node (to which the RTS was sent, and we assume it is not the current node for simplicity) needs to independently receive (detect and decode) the RTS, and the current node needs to receive the CTS, P_{CTS} is

$$P_{CTS} = (P_d P)(P_d P) = k^2 \tag{8.19}$$

The total time during VCS DATA/ACK phase is $BL + t_A$. Thus the expected VCS delay D_{VCS} for each incoming RTS is (the term $t_A/2$ considers the average delay for the successful CTS after successful RTS, when it does arrive)

$$
\begin{aligned}
D_{VCS} &= k[(1 - P_{CTS})t_A + P_{CTS}(BL + t_A + t_A/2)] \\
&= k^3 BL + k(1 + k^2/2)t_A
\end{aligned}
\tag{8.20}
$$

P_Y is the probability of success in the Markov chain represented in Figure 8.5. The expected number of cycles until success is $1/P_Y = N$. Therefore, the restart of RTS contention is expected to happen $1/P_Y - 1 = N - 1$ times. Therefore, W_A can be computed as

$$
\begin{aligned}
W_A &= W_x + (N-1)(W_x + D_{VCS}) \\
&= NW_x + (N-1)[k^3 BL + k(1 + k^2/2)t_A]
\end{aligned}
\tag{8.21}
$$

8.5.3 Expected Time to CTS Reception

We analyze the next sequence until CTS is received for the current node. We use a simple assumption that no collisions happen during the CTS period, assuming RTS-based VCS worked. So CTS loss will only be due to BER and packet detection probability. If the transmitter does not get CTS, it restarts contention window for RTS. Any other node that had received the RTS does a VCS for CTS. It resets and restarts contention if CTS does not arrive. The simplified Markov chain is shown in Figure 8.6. The probability that CTS is received is k^2, which is the detection and decoding probability for both the RTS at the receiver (note that RTS needs to be independently detected and decoded at the receiver, with probability k) and CTS at the transmitter (CTS reception probability is also k). The expected delay in each cycle in the Markov chain is $W_A + t_A$ to include the time until transmission of RTS W_A and time to wait for

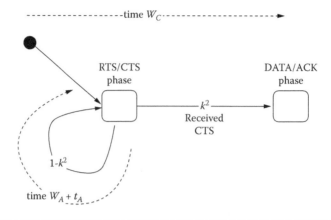

Figure 8.6 Model to compute expected time to CTS reception.

CTS t_A. For simplicity, we consider the expected delay in the last cycle before success also as t_A instead of $t_A/2$. Therefore, the expected time until CTS reception W_c is

$$W_C = \frac{1}{k^2}(W_A + t_A)$$

(8.22)

8.5.4 Expected Throughput

The ACK will be transmitted if any one packet in the train gets through. The probability of getting at least one packet P_{AP} is

$$P_{AP} = 1 - (1-k)^B$$

(8.23)

As $0 < k < 1$, for sufficiently large B, $(1 - k)^B \rightarrow 0$ and hence $P_{AP} \rightarrow 1$. For example, if $P_d = P = 0.9$ and $B = 5$, P_{AP} is 0.9998. Thus, for a sufficiently large number of packets in the batch, we can assume that the receiver will get at least one packet and hence we assume it will send back an ACK with probability 1 at the end of the DATA round.

The probability of successful transmission of any one packet in a batch is k. Thus the expected number of packets delivered in a batch of size B is Bk. The probability of sending an ACK is taken as 1 as mentioned above. The probability of correctly receiving the ACK is k. Thus the expected number of acknowledged packets in every batch transmission is Bk^2.

We can now compute the average throughput of the system. Since the expected time for a single cycle comprising one contention phase followed by a DATA/ACK

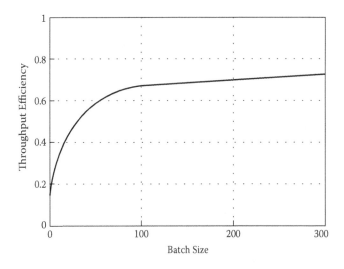

Figure 8.7 **Network throughput vs. batch size. This illustrates that increasing batch size increases the efficiency.**

phase is $W_c + (BL + t_A)$, and the system capacity is $1/L$, the normalized throughput T per node is

$$T = \frac{Bk^2}{W_c + (BL + t_A)} \bigg/ \left(\frac{1}{L}\right)$$

$$= \frac{Bk^4 L}{(W_A + t_A) + k^2(BL + t_A)}$$

(8.24)

This is plotted in Figure 8.7. The parameters used are as follows:

$$L = 0.5,\ N = 3,\ P_d = 0.9,\ P = 0.9,\ D = 0.5,\ W = N$$

This illustrates how increasing batch size improves the throughput. There were some simplifying modeling assumptions made. For example, we assumed that there were no collisions during the CTS sending phase in Section 8.3. During the DATA phase we assumed there were no collisions, but in reality, due to hidden nodes or nodes that missed out RTS and CTS, it's possible. But this simple model captures the essential behavior of a MACA-based protocol and helps the designer to infer characteristics without having to resort to extensive simulations. As mentioned earlier, an alternative analysis by the authors is available.[31]

8.6 Conclusion

In this chapter we compared TDMA, FDMA, and CDMA for use in a distributed topology UAN and showed that TDMA-based schemes are perhaps best suited for the purpose. Dynamic TDMA is a natural extension that addresses the requirement of dynamic channel allocation (ad hoc networks). MACA is seen as a further extension of dynamic TDMA and inherits many of its advantages. It offers even greater robustness as it does not require precise time synchronization, which is often difficult to achieve in many underwater networks. An 802.11-style protocol based on MACA with suitable modifications is perhaps the best choice for general-purpose distributed topology UANs, and has good experimental validation, unlike many other proposals for underwater networks. There are a number of ongoing research projects around the world today exploring suitable variations to the MACA protocol for UANs to improve their performance. In going from pure TDMA to D-TDMA and to MACA, there are trade-offs such as decreased efficiency and increased delay, that come together with benefits such as ad hoc capability, not needing time synchronization, and not having to do re-use planning. The system designer has to choose the trade-offs based on the exact requirements of the network being built.

References

1. J. Heidemann, Y. Wei, J. Wills, A. Syed, and L. Yuan, "Research challenges and applications for underwater sensor networking," in *IEEE Wireless Communications and Networking Conference, WCNC*, 2006, pp. 228-235.
2. K. Jiejun, C. Jun-hong, W. Dapeng, and M. Gerla, "Building underwater ad-hoc networks and sensor networks for large scale real-time aquatic applications," in *IEEE Military Communications Conference, MILCOM*, 2005, vol. 3, pp. 1535-1541.
3. J. Rice, B. Creber, C. Fletcher, P. Baxley, K. Rogers, K. McDonald, D. Rees, M. Wolf, S. Merriam, R. Mehio, J. Proakis, K. Scussel, D. Porta, J. Baker, J. Hardiman, and D. Green, "Evolution of Seaweb underwater acoustic networking," in *OCEANS 2000 MTS/IEEE*, 2000.
4. E. M. Sozer, M. Stojanovic, and J. G. Proakis, "Underwater acoustic networks," *IEEE Journal of Oceanic Engineering*, vol. 25, pp. 72-83, 2000.
5. J. G. Proakis, E. M. Sozer, J. A. Rice, and M. Stojanovic, "Shallow water acoustic networks," *IEEE Communications Magazine*, vol. 39, pp. 114-119, 2001.
6. P. Karn, "MACA—A new channel access method for packet radio," in *ARRL/CRRL Amateur Radio 9th computer Networking Conference*, 1990, pp. 134-140.
7. G. Bianchi, "Performance analysis of the IEEE 802.11 distributed coordination function," *IEEE Journal on Selected Areas in Communications*, vol. 18, pp. 535-547, 2000.
8. T. S. Rappaport, *Wireless Communications, Principles and Practice*, 2 ed. New York: Prentice Hall, 2002.
9. M. Chitre, S. Shahabudeen, and M. Stojanovic, "Underwater acoustic communications and networking: Recent advances and future challenges," *Marine Technology Society Journal—"The State of Technology in 2008,"* vol. 42, Spring 2009 2008.

10. D. Brady and J. A. Catipovic, "Adaptive multiuser detection for underwater acoustical channels," *IEEE Journal of Oceanic Engineering*, vol. 19, pp. 158-165, 1994.
11. A. Kebkal, K. Kebkal, and M. Komar, "Data-link protocol for underwater acoustic networks," in *Oceans 2005—Europe*, 2005, vol. 2, pp. 1174-1180.
12. V. P. Ipatov, *Spread Spectrum and CDMA, Principles and Applications*, New York: John Wiley & Sons Ltd, 2005.
13. S. Shahabudeen, M. Chitre, and M. Motani, "A multi-channel MAC protocol for AUV networks," in *IEEE Oceans '07*, Aberdeen, Scotland, 2007.
14. S. Jarvis, R. Janiesch, K. Fitzpatrick, and R. Morrissey, "Results from recent sea trials of the Underwater Digital Acoustic Telemetry system," in *OCEANS '97. MTS/IEEE Conference Proceedings*, 1997, vol. 1, pp. 702-708.
15. F. Muratore and G. Romano, "GSM versus CDMA: performance comparisons," in *Global Telecommunications Conference, GLOBECOM '96*, 1996, vol. 1, pp. 519-524.
16. P. Bisaglia, F. Boccardi, V. D'Amico, M. Moretti, B. Scanavino, and D. Veronesi, "On the capacity comparison of multi-user access techniques for fourth generation cellular TDD OFDM-based systems," in *Vehicular Technology Conference. VTC 2005-Spring*, 2005, vol. 5, pp. 3077-3081.
17. I. Martoyo, H. Schober, and F. Jondral, "CDMA versus OFDM, a performance comparison in selective fading channels," in *IEEE Seventh International Symposium on Spread Spectrum Techniques and Applications*, 2002, vol. 1, pp. 139-143.
18. M. Chitre, S. H. Ong, and J. Potter, "Performance of coded OFDM in very shallow water channels and snapping shrimp noise," in *MTS/IEEE OCEANS 2005*, 2005, vol. 2, pp. 996-1001.
19. A. Doufexi, S. Armour, M. Butler, A. Nix, D. Bull, J. McGeehan, and P. Karlsson, "A comparison of the HIPERLAN/2 and IEEE 802.11a wireless LAN standards," *IEEE Communications Magazine*, vol. 40, pp. 172-180, 2002.
20. A. P. Dolc and M. Stojanovic, "Optimizing the Transmission Range in an Acoustic Underwater Network," in *OCEANS '07* Vancouver, Canada, 2007.
21. M. Molins and M. Stojanovic, "Slotted FAMA: a MAC protocol for underwater acoustic networks," in *MTS/IEEE OCEANS '06*, 2006.
22. X. Peng and J. H. Cui, "Exploring random access and handshaking techniques in large-scale underwater wireless acoustic sensor networks," in *OCEANS 2006*, 2006, pp. 1-6.
23. J. J. Garcia-Luna-Aceves and C. L. Fullmer, "Performance of floor acquisition multiple access in ad-hoc networks," in *Third IEEE Symposium on Computers and Communications, ISCC '98*, 1998, pp. 63-68.
24. M. Durvy, O. Dousse, and P. Thiran, "Modeling the 802.11 protocol under different capture and sensing capabilities," in *IEEE INFOCOM 2007*, 2007, pp. 2356-2360.
25. X. Guo, M. R. Frater, and M. J. Ryan, "Design of a propagation-delay-tolerant MAC protocol for underwater acoustic sensor networks," *IEEE Journal of Oceanic Engineering*, vol. 34, April 2009.
26. H. Doukkali, L. Nuaymi, and S. Houcke, "Distributed MAC protocols for underwater acoustic data networks," in *IEEE 64th Vehicular Technology Conference, VTC-2006 Fall*, 2006, pp. 1-5.
27. B. Peleato and M. Stojanovic, "A MAC protocol for ad hoc underwater acoustic sensor networks," in *WUWNet '06*, 2006, pp. 113-115.
28. C. Petrioli, R. Petroccia, and M. Stojanovic, "A comparative performance evaluation of MAC protocols for underwater sensor networks," in *OCEANS 2008*, 2008, pp. 1-10.

29. A. Syed, Y. Wei, and J. Heidemann, "Comparison and evaluation of the T-Lohi MAC for underwater acoustic sensor networks," *IEEE Journal on Selected Areas in Communications,* vol. 26, pp. 1731-1743, 2008.

30. R. K. Creber, J. A. Rice, P. A. Baxley, and C. L. A. F. C. L. Fletcher, "Performance of undersea acoustic networking using RTS/CTS handshaking and ARQ retransmission," in *OCEANS, 2001. MTS/IEEE Conference and Exhibition,* 2001, vol. 4, pp. 2083-2086.

31. S. Shahabudeen and M. Motani, "Modeling and performance analysis of MACA based protocols for adhoc underwater networks," in *WUWNet '09* Berkeley, California, 2009.

Chapter 9

Medium Access Control Layer for Underwater Sensor Networks

Yanping Zhang, Yang Xiao, Min Chen,
Praveer Bahri, and Madhulika Kamboj

Contents

There is increasing interest in underwater sensor networks (USNs). Different from terrestrial radio-based sensor networks, communication in underwater sensor networks relies on acoustic signals. Acoustic signals have a propagation speed that is around five orders of magnitude slower than radio signals. Meanwhile, featured by the bandwidth limitations, high transmit energy cost, complex multi-path effects, and high bit-error rates, medium access control (MAC) with an acoustic medium in USNs becomes a complex and challenging problem. In this chapter, we discuss various MAC protocols that are designed for short-range acoustic underwater sensor networks, energy-efficient reliable MAC protocol, and slotted floor acquisition multiple access (FAMA) MAC protocol and low-power acoustic modem for dense underwater sensor networks.

9.1 Introduction

The earth is a water planet. Over two-thirds of the surface of the earth is covered by the ocean [21]. A long time ago, the ocean was a mystery to human beings and thus attracted lots of interest to learn about it. Even in current research, the aqueous environment (including oceans, rivers, lakes, ponds, and reservoirs, etc.) is also a critical research area for many scientists and different applications, such as scientific exploration, commercial exploitation, and attack protection [21].

Usually, underwater environment monitoring is the common way to learn about the ocean. Through monitoring underwater environmental variables, such as water temperature, pressure, conductivity, turbidity, and certain pollutants [22], people have achieved more and more understanding of the ocean. The application of underwater sensor networks has also attracted more and more interest.

The traditional way to monitor the ocean bottom or column is to connect underwater sensors by cables [20], and this approach seriously limits the flexibility and applicability of existing wired underwater surveillance systems. The stability of the system could be threatened by many situations. For instance, wires might be cut off by intruders, which would make monitored data fail to be transported.

Also, a power outage might render the system unusable. Finally, such deployment might be unsuitable for some real-time monitoring applications. For example, in the application of seismic monitoring, the cables connecting sensors or instruments are likely broken due to the rupture of the earth's crust.

To solve the above problem and broaden the underwater surveillance applications, underwater acoustic sensor networks (UASNs) are proposed [20], which rely on the transmission of underwater sensory data through wireless acoustic connections that improve failure resilience otherwise present in their wired counterparts. UASNs can support lots of potential applications, such as monitoring environmental factors (such as seismic event detection in underwater environment, weather condition forecast, etc.) and coordinating marine/submarine equipments (such as navigating ships) [1]. An important example is in the offshore oil industry: UASN enables wireless communication between submarine-controlled vehicles as well as some elements that are above the surface of the sea. The viability of wireless acoustic transmission lowers the cost in installing an oil platform [1] and also improves failure resilience compared with traditional underwater sensor networks.

A UASN [3] is constituted by a number of sensors and underwater vehicles that are deployed in a specific underwater area to perform collaborative monitoring tasks. Sensor nodes and vehicles communicate with each other through acoustic signals [1]. They are required to be capable of self-configuration, which means that they must be able to coordinate to do actions based on exchanging self-configurations like location and movement information and then to relay monitored data to an above-ocean surface station [1].

Underwater sensor networks are different from terrestrial sensor networks in many aspects, including physical, technological, and economic differences. However, research on application-specific protocols of UASNs is still in its early stage. Due to the great difference between the acoustic communication and the terrestrial radio propagation environments, it is hard to say if the experiences for designing the radio signal protocols can be reprocessed in a way so that it can be applicable for underwater acoustic communications [1].

In this chapter, we investigate the features (e.g., more costly equipment, higher mobility, and different energy regimes) of underwater networks, especially related to MAC protocols. Although there has been some work on the development of MAC [10] and routing protocols, the fundamental networking primitive which is broadcast, has not been fully explored yet. Broadcast is the basic and essential way for varieties of vital networking functions, like neighbor discovery, route establishment, and data transmissions [1]. Broadcast also has a higher probability of correct delivery than directly routing to destination in some specific applications like tsunami detection [1]. Meanwhile, reliable broadcast is required in some network applications such as network reprogramming of nodes [1]. However, for the higher costs for acoustic modems, the broadcast among underwater acoustic sensors is still unreliable [1]. Thus, the unique properties of the underwater acoustic channel should be furnished to design novel protocols with reliable broadcast that are different from radio networks.

The rest of the chapter is organized as follows. In Section 9.2, we present the basic practical issues in UASNs. Then, we discuss various challenges of MAC protocol design for UASNs in Section 9.3. We will study different MAC and routing protocols in UASNs in Sections 9.4 and 9.5, respectively.

9.2. UASN Communication Architecture

9.2.1 Two Types of Architecture

In [20], two types of architectures are discussed: static two-dimensional UASNs for ocean bottom monitoring, and static three-dimensional UASNs for ocean column monitoring.

9.2.1.1 Two-Dimensional UASNs

In two-dimensional architecture, sensor nodes are anchored to the bottom of the monitored underwater environment, such as an ocean, river, etc. [20]. The wireless acoustic links of UWSNs (Underwater Sensor Network) are established based on the interconnection between sensor nodes with one or more underwater sinks, which are also called UW-sinks, functioning to relay data from the underwater sensors to the surface stations [20]. For this purpose, two acoustic transceivers are required on UW-sinks [20]. One is a vertical transceiver and the other is a horizontal transceiver [20]. The horizontal transceiver is used to communicate with the sensor nodes in terms of UW-sink to sensors communication and sensors to UW-sink communication, respectively. By comparison, the vertical transceiver of UW-sink relays data to a surface station. The surface station is capable of handling multiple parallel acoustic signals sent by the surrounding UW-sinks. In addition, a radio frequency (RF)/satellite transmitter with long range is also deployed in the surface station to communicate with the onshore sink.

The connection between sensor and UW-sink can be established in two ways: direct link or multi-hop routing. By using direct link, sensors are able to send data directly to the selected UW-sink, while in multi-hop routing approach the gathered data is forwarded by multiple intermediate nodes until it reaches the UW-sink [20].

9.2.1.2 Three-Dimensional UASNs

In three-dimensional (3D) architecture, sensor nodes are deployed to be floating at different depths in the ocean for cooperative 3D environmental sampling. Meanwhile, each UW-sensor node is fixed to a surface buoy by a wire that can be adjusted for proper and suitable length, which is the depth of the sensor [20]. However, this architecture faces many challenges. First, sensors need to collaboratively regulate their depths to maximize the coverage of the whole network according to sensing ranges and communication coverage. Second, since there is no UW-sink deployed underwater, collected data should be correctly relayed to the surface station via multiple hops. Therefore, network devices should coordinate their depths so that for each sensor there should be at least one existing path for it to be connected with at least one surface station [20].

9.2.2 Underwater Network Operating Regime

Spatial coverage and node density are used to characterize underwater networks. Significant implicit factors related to the MAC layer and the network layer must be considered as design issues [1].

Two different scales of communications should be considered according to acoustic range of the nodes. For nodes that are in direct contact, the network works as a single-hop network, which can be centralized or distributed control. Furthermore, for multiple-hop communication, the network is a larger communication network for data to reach destination. There is a situation that geographic coverage might be larger than the unpartitioned link-layer coverage of all nodes and then disruption-tolerant networking (DTN) routing techniques are required [1]. Due to the unique feature of acoustic signals, when lots of sensor nodes are deployed in a small area, conflicts will be a serious problem for communication [1]. Second, it will cost a lot for dense deployment in a huge underwater environment, which makes DTNs an attractive solution [1]. In a UASN, both single-hop and multi-hop clusters can be deployed to construct the whole network, in which DTN routing could be employed for infrequent communication [1].

Catipovic described the features of underwater acoustic channels in [2]. He also reviews the recent work implemented with another two media in underwater networks, i.e., long-wave radio and optical underwater networks. He also explains related technological limits for nodes and the further influence on the network topology, etc.

9.2.2.1 Physical Channel

In the underwater environment, acoustic signals are the main way used for communication. Neither radio signal nor optical signal is appropriate for underwater communication. Ocean water is very salty, which critically attenuates the radio waves [1]. There are also some applications of long-wave radio, but they are only applicable for short-distance communication. Light signal is easily scattered and absorbed by water, although some connections in absolutely clear water working with short range and high bandwidth may employ blue-green wavelengths [1]. In the underwater environment, optical signal is also considered as an efficient communication media only for low-cost, short-range connections of order 1–2 m [1]. The expectation of date rate for optical modems in extremely clear water is several

Mbits/sec at ranges up to 100 m [1]. Therefore, acoustic signal is the only appropriate way for communication for long-range communication in the underwater environment with common water clarity. The typical feature accompanied by acoustic communication is great propagation delay due to the slow spread speed of sound in water, which is approximately 1500 m/sec, five orders of magnitude lower than the speed of light [1]. Compared to radio signal, acoustic signal has several other aspects of constraints, such as correctness, bandwidth, and channel dependency [1]. First, there is a higher bit-error probability in acoustic communication because of its phase and amplitude fluctuations, while forward error correction or error correction coding is required by radio channels [1]. Second, the bandwidth of acoustic communication is quite limited due to the strong attenuation, especially with increasing frequency [1]. Third, acoustic communication can be disturbed by the environment, and the most common disturbance is multipath interference, which causes frequency-selectivity of the channel. Such frequency-dependent interference is always time-varying and might be caused by many different factors such as surface waves or vehicle motion [1]. However, the propagation delay of acoustic channels can always be estimated and stable enough for configuring the network protocols [1].

9.2.2.2 Technological Limitations

The communication of an underwater acoustic network is always half-duplex. The acoustic transducers can only do one thing of transmission and reception at the same time [1]. Because of the space constraints of the underwater environment, the network cannot provide far enough space for transducers in different frequencies for establishing full-duplex connections [1]. Both autonomous underwater vehicles (AUVs) and compact stationary nodes follow the constraint. Meanwhile, the transducer size is proportional to wavelength and usually only higher center frequencies are available for small AUVs [1]. Furthermore, small AUVs can transmit data at high rates while they cannot receive data at such high rates. There are mainly two aspects that contribute to the asymmetry: propulsion noise and mounting receiver of small AUVs [1]. The asymmetry in sending and receiving rates is also the main reason of the popularity of star topologies with base stations in current mobile underwater networks [1].

9.3 Challenges of MAC Protocol Design for UASNs

Because of the unique characteristics of acoustic channel and propagation, the design of acoustic communication sensor networks is a difficult problem. In this section, underwater acoustic communication channel and associated MAC layer challenges for underwater networking are summarized. The design challenges for UASNs are also discussed.

9.3.1 Underwater Communication Channels

The spread of acoustic signal in underwater environment is about 1500 m/sec, which is five orders of magnitude lower than radio propagation speed. Only very limited bandwidth of underwater acoustic channels is available, which could be influenced by many factors, such as transmission range, frequency, etc. [21].

9.3.2 The Impact of Acoustic Propagation

The challenge of an underwater acoustic communication system is mainly caused by acoustic propagation in underwater environment. In [23] several aspects were discussed, such as speed of sound, channel latency, ambient noise, etc.

- *Acoustic signal propagation in seawater*: Spreading loss and absorption loss are critical features of acoustic propagation [23]. When acoustic signal is sent out, the energy of the signal is fixed and expands when the signal is transmitted over a large surface area [23]. Usually, sphere is used to describe the surface area, especially for short ranges [23], and the decay of the signal energy is at a rate of R^{-2} where R is the distance from source [23]. Meanwhile, the surface and seabed form a natural boundary of the underwater environment, which also bound the range of acoustic communication [23]. Sometimes, when acoustic signals are sent out from a source, the signal cannot vertically spread [23]. Meanwhile, the spread of the signal, which should be spherical spreading, may change to cylindrical spreading. Such a situation may occur especially when the ranges are larger than the depth of the water [23]. Therefore, the loss is the energy conversion during the propagation into heat, which is called absorption loss [23].
- *Waveguide propagation, multipath, and shadow zones*: Acoustic signals can be refracted and reflected due to the environment. The refraction happens because the speed of sound varies spatially in the water column, while the reflection happens because of the bound formed by sea surface and bottom [23]. Both refractions and reflections could result in that the signal propagates in multiple paths to the destination, which could result in inter-symbol interference at the receiver [23]. Meanwhile, temporal fluctuations have great relationship with the propagation environment variations and transmitting or receiving platforms [23].
- *Scattering surface*: The moving sea surface can make the transmitted signal be scattered, which is a seriously challenging communication scenario [23]. Rough sea surface provides various delays of surface bounce paths and reduces the space connection of scattered signals, which results in channel impulse response with high intensity [23].
- *Bubbles*: Breaking waves at the sea surface can produce bubbles, which greatly influence the propagation of high frequency acoustic signals both in open ocean

and near shore regions [23]. Meanwhile, different layers of bubbles existing near the surface can cause a significant attenuation of surface scattered signals [23].

■ *Environment noise*: Some natural sources, such as biological sources and rain, leading to environment noise in the ocean are breaking waves and bubbles. The common theme known for ambient noise is that for higher frequency, there will be a decrease of the power spectral density of the noise [23].

9.3.3 Considerations for the Design of Underwater Protocols

Lots of factors are able to hold great impact on the communication of underwater acoustic signals, such as transmission loss, noise, multipath, Doppler spread, varieties of propagation delays, as well as availability of range and frequency-dependent bandwidth [24]. The bandwidth available for long-range systems, such as over tens of kilometers, is only a few kHz, while the bandwidth for short-range systems, such as operating over only several tens of meters, is more than a hundred kHz [24].

The depth of water is a serious factor impacting UANs. Shallow water refers to water with depth lower than 100 m, while deep water is used for deeper oceans. There are many factors that influence underwater acoustic communication, as follows:

■ *Transmission loss*: Attenuation and geometric spreading are the main concerns of transmission loss [24]. The attenuation mainly refers to the energy absorption or conversion into heat. Larger distance or higher frequency corresponds to more serious attenuation [24]. The geometric spreading can also spread the energy of acoustic signals because of the expansions of the wave fronts. Propagation distance could increase the geometric spreading, while frequency of the signal has nothing to do with it.

■ *Noise, including man-made noise and ambient noise*: Environment noise includes man made noise and natural noise. Man-made noises mainly refer to machinery noise like pumps, reduction gears, and shipping activity [24]. Natural phenomena like hydrodynamics, seismic, and biological phenomena can cause ambient noise [24].

■ *Multipath*: The propagation in multipath can severely degrade the acoustic signal. The link configuration such as horizontal channel characterization determines the geometry of multipath [24].

■ *High delay and delay variance*: The speed of sound in the underwater environment is five orders of magnitude lower than the radio signals. The throughput of the system can be reduced considerably by large propagation delay and its high variance.

■ *Doppler spread*: The Doppler frequency spread can highly degrade digital communications and transmissions, which leads to the interference of many adjacent symbols at the receiver [18]. Some other situations are also greatly related to the Doppler spreading, such as simple frequency translation and a continuous spread of frequencies [24].

- *Bandwidth*: Bandwidth is very much limited in ranges of 0.1 Km and 1000 Km to nearly 100 kHz and 1 kHz, respectively, leading fairly lower data rates than terrestrial wireless communications [13].
- *Attenuation*: The channel of underwater acoustic signal is quite impaired, resulting from absorption, multipath, attenuation, and fading problems. Specifically, there are significant absorptive losses in underwater acoustic signals, which greatly depend on frequency [13].
- *Shadow zones and channel characteristics*: The extreme characteristics of the underwater channel like salinity, density, and temperature variations may cause high bit error and disconnection [13].

9.4 MAC Protocols for Underwater Acoustic Sensor Networks

Medium access is an open problem in underwater acoustic networks [1]. In traditional radio networks, MAC protocols are of great interest in radio-based sensor networks and have been studied for at least 10 years.

9.4.1 Recent Work in Underwater MAC Protocols

A variety of MAC protocols in underwater networks have been explored. Some MAC protocols like Aloha [14] have already been discussed in various research papers. In order to avoid collisions in the situation when two stations transmit concurrently, carrier sensing multiple access (CSMA) [15] and its mutants have been employed. Based on such protocols, stations are required to "listen" to the channel before transmitting packets, and after they are sure there won't be any collision they can continue to transmit. In fully connected and small propagation networks, CSMA is an efficient protocol. However, for the situation of both hidden and exposed terminal problems, it is insecure to employ protocols based on CSMA [1].

In [16], a MAC protocol named MACA (Multiple Access Collision Avoidance) is proposed. In this protocol, any node planning to transmit data will send a control packet first to the destination node, which is called RTS (Request To Send) [13]. Then the destination node replies the sender a CTS (Clear To Send) control packet, which also warns all its neighbors about its activity of sending packets. In [17], Bharghavan proposed a protocol called MACAW (MACA-Wireless), which modified some parts of the original MACA protocol. An adaptive back-off algorithm is exploited in this protocol [17]. And an ARQ (Automatic Repeat-Request) technique is also added as a feature of this protocol, which resends packets with errors [17]. Fullmer and Garcia-Luna-Aceves [4] described the conditions to avoid collisions among data packets in a MACA. In [4], FAMA is exploited to prevent data packets from collisions by extending the time slots for the RTS and CTS. All these protocols need multi-way handshakes so propagation delays seriously reduce their efficiency. In order to make MACA, MACAW, and FAMA applicable for UANs, researchers have spent lots of effort and made related adaptations to these protocols. In [13], Molins and Stojanovic proposed Slotted FAMA, which lessens the impact of propagation delays by introducing timeslots to FAMA. In [6], Kebkal et al. suggested the introduction of an acknowledgment (ACK) scheme in data transmission in order to reduce the impact of propagation delay on FAMA- and MACAW-based protocols, as well as applying code division multiple access (CDMA) for RTS packets to avoid collisions. Foo et al. [7] provide a more detailed proposition to extend CDMA to MACA with references to the radio-based MAC protocols.

There are also efforts, like current Seaweb implementations, trying to use combined TDMA/CDMA clusters, which is also described by Salva-Garau and Stojanovic [8]. This approach shortens the TDMA slot while increasing overheads and the interference probability among clusters. In [9], Doukkali and Nuaymi analyze several MAC protocols for the underwater environment, including the TDMA-CDMA clusters [9].

Energy efficiency is a critical consideration not only in terrestrial sensor networks but also in underwater networks. Coordinated-sleeping MAC protocols such as S-MAC are the result of efforts to overcome energy constraints in terrestrial sensor networks [26]. Adopting these ideas as well as referencing some other MAC protocols both in underwater and terrestrial networks, Park and Rodoplu [5] proposed an energy-efficient UWAN-MAC protocol for delay-tolerant underwater sensor networks. In the following discussion, we introduce some MAC protocols designed for underwater acoustic networks.

9.4.2 Tone Lohi MAC Protocol [12]

In [12], Tone Lohi (T-Lohi) is discussed, which is a reservation-based MAC protocol. T-Lohi MAC is proposed with mainly two ideas. One idea is to detect and count the number of contenders during the reservation, and a traffic-adaptive back-off algorithm is proposed according to that number. As the duration for the occupication of the channel by contention packets is much less than the propagation delay, nodes are able to detect and count contenders as long as the duration of the occupication [12]. The other idea is employing a wake-up tone for the purpose of reserving the data transmission. Each node is equipped with a hardware wake-up tone detector on the acoustic modem, based on which nodes can spend minimal energy during listening to the tone. Substantial energy savings can be achieved by applying the wake-up tone during the reservation phase. The reservation protects the data transmission from collision and saves energy [12].

A reservation period is included in the T-Lohi protocol. The period includes multiple contention slots and is typically on the order of tenths of a second, and after the period is a data transmission period. When a node tries to reserve the data transmission period, it will first transmit a tone during the reservation period. Other nodes will back off when they hear the tone. And a node can finally occupy the data period to transmit data only if it successfully finished the transmission of the tone without hearing another tone during a contention slot [12]. T-Lohi protocol has mainly two variants: synchronized tone (ST) and unsynchronized tone (UT) Lohi protocols. For brevity we only describe the more efficient protocol, ST-Lohi [12].

In ST-Lohi [6], all stations are aligned to contention slots, i.e., a length of the maximum propagation delay and the tone length. Each station sends reservation tones at the beginning of these slots, if not restricted by back-off. During the rest of the contention slot after finishing the tone, the node will wait and listen to possible arrival tones. If no tones are heard, it wins the reservation, and immediately sends its data, as shown in Figure 9.1.

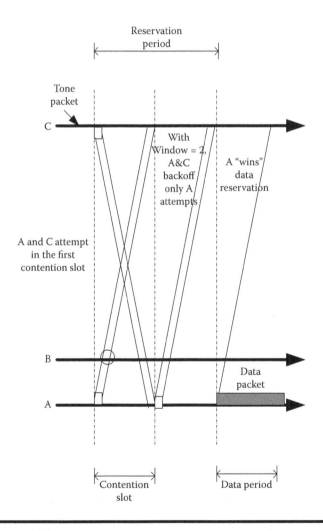

Figure 9.1   ST-Lohi MAC.

If multiple nodes try to reserve the medium at the same time, they have to back off and retry later. The back-off nodes will count the tones received in a given slot, that is, the number of contenders, which is used as their back-off window size. After the data packet is finally transmitted, nodes with previous contention do so with a smaller window compared to those nodes without contention [12].

An interesting observation of medium access in acoustic medium with high latency is space-time uncertainty [12]. Not only concurrent transmissions but also transmissions at different time and distances can cause collisions. To synchronize the transmission time is to remove one dimension of uncertainty, and to wait for the maximum propagation delay can remove the other [12]. This uncertainty is due to the scheme of ST-Lohi, waiting for enough time (the maximum propagation time) in order to detect any possible collisions [12]. When there are no tones of contention, the node successfully "wins" the reservation of the channel and its data can be transmitted in the next time slot. Usually, nodes are at different physical places in space, so even though they might transmit at the same time, their contention tones arrive at a different time [12]. Nodes are allowed to count the number of contenders based on the space-time separation and the count is a basic consideration to intelligently select the back-off for subsequent contention periods. Using the wake-up tone abstraction can achieve energy efficiency, which means that nodes don't have to be fully active for the entire contention slot to transmit/receive tones [12].

9.4.3 Slotted FAMA

As FAMA [4] protocol needs RTS and CTS packets, it would not be efficient in underwater acoustic networks. And data collisions are another problem in FAMA when violating these conditions, which is shown in Figure 9.2. The problem can be solved if the new packets are forbidden to be sent while data is being transmitted. Nodes should be restricted to send any packet if it would collide with a current

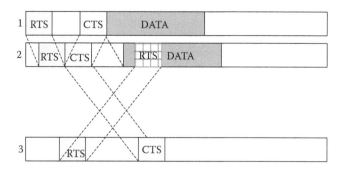

Figure 9.2 RTS from C collides with data packet from A. (After M. Molins and M. Stojanovic, "Slotted FAMA: a MAC protocol for underwater acoustic networks." In *Proceedings of the IEEE OCEANS '06*.)

transmission. Slotted FAMA [13] applies a restriction on the time when packets are sent, which slots the time to eliminate the asynchronous nature of the protocols. All kinds of packets, including RTS, CTS, DATA, or ACK, must be transmitted at the beginning of the slot. The way to choose the length of the slot is to avoid all packet collisions. It seems that all nodes should know whether they will collide with an ongoing transmission if they send a packet at the beginning of the next time slot [13]. The way to accompany this is to set the time slot to be the total value of the maximum propagation delay τ and the transmission time of a CTS packet γ [13]. Such a choice of the time slot is to make sure that there is enough time for all the nodes within range to receive the RTS or CTS packet [13].

In Slotted FAMA [13], when a node needs to send a packet it first transmits an RTS packet and waits for the next slot, including the destination that will receive the packet within the slot time. The destination node will then send a CTS packet at the beginning of the next slot [13]. All the terminals within the transition's range of the destination including the source node receive this CTS packet within the slot time. After receiving the CTS, the source knows that it has won the channel to send data, and it waits to send its data packet at the beginning of the next slot. After receiving the data packet, the receiver will also send an ACK packet to confirm the successful transmission. Figure 9.3 shows the whole process [13].

Slotted FAMA [13] is mainly based on carrier sensing in which terminals are always listening to the channel. Terminals are keeping in idle state when they don't have packets to transmit or they cannot sense the carrier in the channel. When a node has a packet to send and doesn't detect any carrier, the terminal sends an RTS and waits two slots for a CTS packet. During this time if no CTS is received, the terminal assumes a collision existed and turns to back-off state for several slots, which is randomly decided. During the back-off period if no carrier is sensed the terminal re-sends the RTS packet. And until the terminal successfully receives CTS, it will begin transmitting the packet in the next time slot [13].

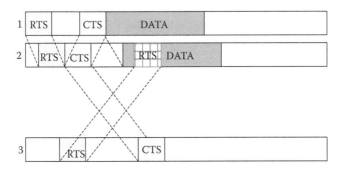

Figure 9.3   A successful handshake between terminals A and B in Slotted FAMA. (After M. Molins and M. Stojanovic, "Slotted FAMA: a MAC protocol for underwater acoustic networks." In *Proceedings of the IEEE OCEANS '06.***)**

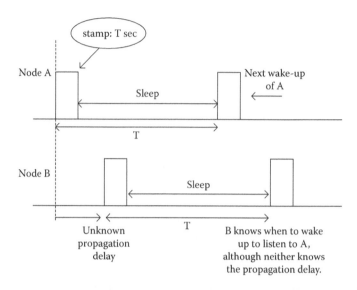

Figure 9.4   Basic idea of MAC protocol. (After V. Rodoplu and M. K. Park, "An energy efficient MAC protocol for underwater wireless acoustic networks." In *Proc. IEEE Oceans Conf.,* **vol. 2, pp. 1198-1203, Sept. 2005.)**

9.4.4 An Energy-Efficient MAC Protocol for Underwater Wireless Acoustic Networks

An energy-efficient distributed and scalable MAC protocol is presented in [5] that can work in spite of long, unknown propagation delays of acoustic medium in the underwater environment. This energy-efficient MAC protocol [5] can be used for underwater acoustic networks in which sensor nodes are quite energy-limited. Energy rather than bandwidth utilization is the main performance metric of such MAC protocols, which significantly differentiate the energy-efficient protocol from ALOHA, MACA, and MACAW protocols [5]. The basic idea of the energy-efficient protocol [5] is as shown in Figure 9.4.

As a preamble, node A will send a beacon signal before its data transmission in any transmission cycle. The transmission cycle "T" of node A is announced. Whenever a new node (node B) tries to join the network it achieves frame synchronization by first listening to the channel for this preamble sequence [5].

Node B can achieve the value of transmission cycle period (T) through decoding the beacon message. This scheme stamping the transmission cycle explicitly enables node B to wake up at exactly the correct time in the next cycle. As the propagation delay from one node to another normally remains the same, node B even has no need to know the value of the propagation delay [5]. This type of localized protocol can hold for any two nodes (A, B) [5]. Two types of collisions may occur in this protocol, which are "receive–receive collision" and "transmit–receive collision" [5]. "Receive–receive collision" can occur when a node concurrently receives more than two data packets that overlap in duration. In such a situation both of the two packets cannot be decoded correctly and related information is lost. "Transmit-receive collision" refers to when a node is transmitting a data packet and another packet from other nodes arrives at the node and collides with the node's own transmission [5].

The topology control layer in [5] monitors the transmission of a node to track its neighbors and then determines when to wake up which node. The above scheme is used to predetermine the listen times and each node initially transmits a packet randomly and independently. In the protocol, once a transmission start time for a node is chosen, the node will send its data packet in the next cycle according to the schedule [5].

This protocol enhances the existing protocol by choosing the duration of listen time. The duration is chosen as the sum of two values: SYNC packet duration and the maximum propagation delay between any two neighbors. Meanwhile, the choice of the length of duration induces trade-off. If the duration is too long, it induces energy efficiency of the protocol, while if it is too short it might lose some messages from the newcomers [5]. Also, in [5], guard time is adopted for both sides of its transmission durations.

9.4.5 Modified MAC Protocol Design for Underwater Acoustic Data Communication [19]

There are some limitations to MAC protocol for UASNs including the acoustic transmission signal discussed before. The traffic data for an image transmission mainly requires the bit rate to be about 10–50 kbps. However, the physical layer of UASN, that is, the acoustical transducers, can support only 1–10 kbps bit rate [19]. When multi-media data packets need to be transmitted in UASN, one channel of acoustic medium is not capable enough to accomplish the transmissions. One way to extend the bit rate is to utilize multiple channels with different frequencies [19]. The protocol designed in [19] tries to balance the throughput and expected delay, both of which are mostly dependant on the transmission media and the transceiver [19].

There are many stations as well as many available acoustic channels in the underwater network. And each station can choose one of these channels to transmit data. In the following paragraphs, we study the collision-free protocol designed in [19].

Figure 9.5 shows a carrier-sensing transition algorithm, used when overflow is caused by the increasing flag [19]. Overflow can accumulate the idle time for each channel [19]. If the carrier sensing is detected, the flags for the related channel are set to 0 and the timer operates. The carrier sensing for the transmitting channel is detected as bit unit [19].

Figure 9.6 describes the receiving state transition diagram. After receiving a frame, a station will first send ACK to related station. And then the station again turns to carrier sensing mode and processes the received frame. If the frame is with

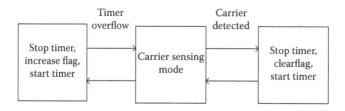

Figure 9.5 The carrier-sensing static diagram. (J. K. Yeo, Y. Lim, and H. H. Lee, "Modified MAC protocol design for the acoustic-based underwater data communication," *Proc. of IEEE ISIE 2001.***)**

some error, the station will send NAK (Negative Acknowledgment) signal to related stations and return to carrier sensing mode as well as discard the received frame.

The transmitting frame is built by requesting to frame transmit signal [19]. Figure 9.7 shows the transition procedure. In the transmitting frame one acoustical channel among the channels should be selected. In the first stage the candidate channel transmits the frame in its own station and destination station needs to receivable channels. Figure 9.8 shows the searching procedure and Figure 9.9 shows the selecting procedure of candidate channels. The channel with the longest idle time is selected to transmit frame and after the transmission it returns to ACK receiving mode. In the case that no idle channel is found in the procedure, in order to search for candidate channels, carrier sensing is made [19].

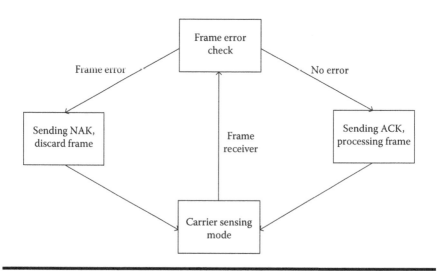

Figure 9.6 The receiving state transition diagram. (J. K. Yeo, Y. Lim, and H. H. Lee, "Modified MAC protocol design for the acoustic-based underwater data communication," *Proc. of IEEE ISIE 2001.***)**

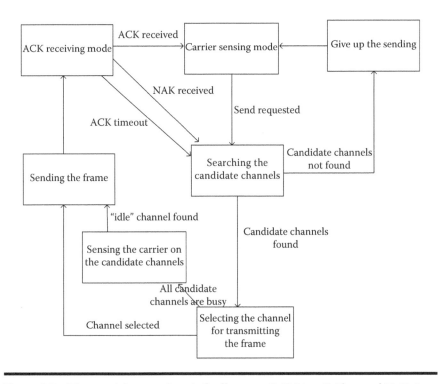

Figure 9.7   The receiving sensing static diagram. (J. K. Yeo, Y. Lim, and H. H. Lee, "Modified MAC protocol design for the acoustic-based underwater data communication," *Proc. of IEEE ISIE 2001.***)**

Figure 9.10 demonstrates the error control method, which includes the activity of stopping and waiting for ARQ. After finishing a frame transmission, the station turns back to ACK receive mode. If receiving NAK or not receiving ACK successfully, the station has to retransmit the frame, just following the procedures previously explained [19].

9.5 Routing Protocols for Underwater Acoustic Sensor Network

9.5.1 Three-Dimensional Routing Algorithms for Delay-Insensitive and Delay-Sensitive Applications in Underwater Sensor Networks

In both [11] and [25], the features of underwater acoustic channels are discussed. Gathering data is an important consideration in a network, and that's also why routing protocols are of such great concern to many researchers. In [25], cross-layer

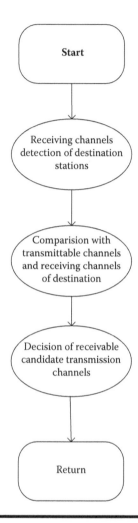

Figure 9.8 The procedure for searching candidate channels.

based network and layer data gathering are discussed. This paper also discussed the interactions between the routing and the unique features of underwater acoustic medium. In addition, the distributed delay-insensitive routing algorithms and delay-sensitive routing algorithms are also presented in the above papers. According to different application requirements and varying conditions of underwater acoustic channels, nodes can choose the next hop node for saving energy consumption [11].

With the purpose of minimizing energy consumption in the network, authors in [25] proposed two kinds of routing algorithms for delay-insensitive and delay-sensitive sensor network applications. Both algorithms allow sensor nodes to choose their next hop for an efficient path to save energy [25].

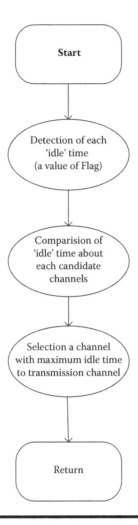

Figure 9.9   The procedure for selecting the sending channel. (J. K. Yeo, Y. Lim, and H. H. Lee, "Modified MAC protocol design for the acoustic-based underwater data communication," *Proc. of IEEE ISIE 2001.***)**

Compared to terrestrial sensor networks, underwater acoustic sensor networks are featured with more challenges, especially because of the more complex underwater environment and the larger delay of acoustic communication channels [25]. The most significant considerations of underwater acoustic sensor networks are summarized as follows [25]: (1) Compared to radio-based communication in terrestrial sensor networks, the speed of acoustic signal in the underwater environment is five orders of magnitude slower, which leads to much longer propagation delay for communication in UASNs. (2) In the underwater environment, acoustic signals have multipath and fading problems during the transmission, which

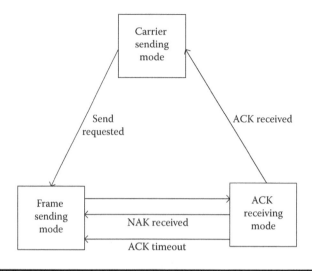

Figure 9.10 The state transition diagram for receiving ACK. (J. K. Yeo, Y. Lim, and H. H. Lee, "Modified MAC protocol design for the acoustic-based underwater data communication," *Proc. of IEEE ISIE 2001.***)**

make the underwater acoustic channel severely impaired. (3) Acoustic channels are always corresponded to high bit error rates as well as vulnerable connectivity, such as shadow zone. (4) The bandwidth of acoustic medium is quite limited, which severely impacts the information transmission. (5) Usually, after deployment in an underwater environment, sensor nodes are not easy to recharge so that their battery power is quite limited. (6) Sensors may be eroded or foul in water, which makes the sensor nodes unable to work properly.

Delay-Insensitive Routing: In [25], delay insensitive routing in the 3D underwater environment is proposed for delay-insensitive applications. This routing algorithm is a distributed geographical routing solution. With the purpose of exploiting the channel efficiently as well as minimizing energy consumption, the algorithm proposes the concept of packet train [25]. The idea of packet train is very similar to the idea in IEEE 802.11 [27–38]. Packet train is defined as juxtaposition of packets and it is a way of transmission such that packets are transmitted back-to-back by a node and the node does not need to release the channel in a single atomic transmission. The ACK scheme is also included in the algorithm, which may need some packets to be retransmitted if ACKs are not correctly received in time. And in this algorithm, there are two kinds of retransmissions. One is to selectively request specific packets included in the next train to be retransmitted, and the other is to cumulatively retransmit all packets that are included in the whole train.

From the energy-saving point of view, the author designs the algorithm to allow nodes to select their next hop based on minimizing the energy cost [25]. When exploiting links, the algorithm tries to find those with low error rates in packets, that is, to find a way with maximum possibility that the receiver can correctly decode the data packet [25]. Meanwhile, the energy efficiency of the path is weighted by the retransmission number that is required during the communication, whose objective it is to save energy [25].

The proposed strategy in [25] can achieve two objectives. First, it can promote the efficiency of the channel that is achieved through the increment of the size of the transmitted train. Second, due to the short length of data packets, the error rate of transmissions is limited. The length of data packets greatly determines the error rate of the packets. The approach in this algorithm decouples the impact of the length of data packets and the choice of the train size [25]. Normally, when there is a need, the size of the transmitted train can be increased, which further increases the utilization of the channel [25].

Delay-Sensitive Routing: There are already many efforts spent on the development of the routing protocol in terrestrial sensor networks. For terrestrial ad-hoc and sensor networks, when designing routing protocols, researchers mainly follow a packet switching paradigm. In such a paradigm, the routing functions work separately on each single packet and all the paths used during the routing are dynamically established [25]. UASNs are quite different from the terrestrial sensor network, which requires applicable routing protocols in the underwater environment. In UASNs, information or data packet is transmitted through acoustic medium, which is accompanied by large propagation delay and vulnerability in underwater environment. Meanwhile, compared to terrestrial networks, UASNs are relatively scarce networks because it is not as easy and convenient to deploy many sensor nodes as on land [25]. Therefore, centralized planning of network topology is more applicable for UASNs, which can optimally utilize the resources in the network [25]. The design of a routing protocol for UASNs can be achieved by devising some centralized protocols [25]. Because of the reasons discussed above, one technique that could be considered for a delay-sensitive application in UASNs is virtual circuit routing technique. In such a technique, between each source and sink the multi-hop connections are established *a priori* and packets are also associated with a specific path [25]. For the technique we discussed here, a centralized coordination is needed, which will limit the flexibility of the network architecture [25]. The strong point of the technique is that with a centralized station the network can be exploited to achieve optimal performance at the network layer. For example, with the knowledge of global information, the central station is able to find a path according to needs, such as minimum delay paths for a time-urgent packet or energy-efficient paths just for saving energy within the network, etc. In [25], the authors concluded the problems related to the design of a 3D routing protocol for delay-sensitive applications. The main optimization idea is to find two multi-hop data paths from each source to the station on the surface [25], which are called primary and backup data paths. Under such a scheme, if the primary path fails or some nodes on it cannot work normally, the backup path can offer protection for the communication [25].

9.6 Conclusion

In this chapter we have summarized practical issues of the differences between acoustic-based underwater networks and terrestrial sensor networks that are based on radio communication. The acoustic communication medium is one important factor that distinguishes underwater sensor networks from terrestrial sensor networks. We first studied the features of the acoustic communication channel, especially its propagation character in the underwater environment. Then we analyzed the challenges of designing a MAC protocol for UASNs. After this analysis, we studied current achievements on MAC protocols for UASNs. Several MAC protocols for UASNs were introduced, which includes ST-Lohi MAC protocol, a reservation-based MAC protocol, as well as the Slotted FAMA, which reduces the collisions among data packets and doesn't even require the size of data packets. In Slotted FAMA a simple back-off scheme is employed to achieve the avoidance of collisions. Power control was the main issue studied. Finally, some routing protocols were presented.

Acknowledgment

This work is supported in part by the U.S. National Science Foundation (NSF) under the grant numbers CCF-0829827, CNS-0716211, and CNS-0737325.

References

1. J. Partan, J. Kurose, and B. N. Levine, "A survey of practical issues in underwater networks.", In *WUWNet '06*.
2. J. Catipovic, "Performance limitations in underwater acoustic telemetry," *IEEE J. Oceanic Eng.*, Vol. 15, pp. 205–216, Jul. 1990.
3. M. Stojanovic, "Recent advances in high-speed underwater acoustic communication," *IEEE J. Oceanic Eng.*, 25(2):125–136, Apr. 1996.
4. C.L. Fullmer and J.J.Garcia-Luna-Aceves, "Floor acquisition multiple access (FAMA) for packet-radio networks." In *SIGCOMM' 95*.
5. V. Rodoplu and M. K. Park, "An energy efficient MAC Protocol for Underwater Wireless Acoustic Networks." In *Proc. IEEE Oceans Conf.*, Vol. 2, pp. 1198–1203, Sept. 2005.
6. A. Kebkal, K. Kebkal, and M. Komar, "Data-link protocol for underwater acoustic networks." In *Proc. IEEE Oceans Europe 2005*, pp. 1174–1180, 2005.
7. K. Foo, P. Atkins, T. Collins, C. Morley, and J. Davies, "A routing and channel-access approach for an ad hoc underwater acoustic network." In *Proc. IEEE Oceans 2004*, Vol. 2, pp. 789–795, Nov. 2004.
8. F. Salva-Garau and M. Stojanovic, "Multi-cluster protocol for ad hoc mobile Underwater Acoustic Networks." In *Proc. IEEE OCEANS'03 Conf.*, Sept. 2003.
9. H. Doukkali and L. Nuaymi, "Analysis of MAC protocols for underwater Acoustic Data Networks." In *Proc. IEEE Vehicular Tech. Conf.*, May 2005.

10. W. Ye, J. Heidemann, and D. Estrin, "Medium access control with coordinated adaptive sleeping for wireless sensor networks," *IEEE/ACM Transactions on Networking*, *2004*, 12(4):493–506.
11. D. Pompili, T. Melodia, and I. F. Akyildiz, "Routing algorithms for delay-insensitive and delay-sensitive applications in underwater sensor networks." In *Proc. of ACM Mobicom*, 2006.
12. A. Syed, W. Ye, and J. Heidemann, "Medium access for underwater acoustic sensor networks," USC/ISI technical report ISI-TR-624, October 2006.
13. M. Molins and M. Stojanovic, "Slotted FAMA: a MAC protocol for underwater acoustic networks." In *Proceedings of the IEEE OCEANS '06*.
14. N. Abramson, "The aloha system." In *Computer-Communication Networks*, Englewood Cliffs, New Jersey: Prentice Hall, 1973.
15. L. Kleinrock and F. A. Tobagi, "Packet switching in raio channels: part I carrier sense multiple-access modes and their throughput-delay characteristics," *IEEE Trans. On Commun.*, Vol. COMM-23.
16. P. Karn, "MACA—A new channel access method for packet radio," ARRL/CRRL Amateur Radio 9th Computer Networking Conference, 1990.
17. V. Bharghavan, A. Demers, S. Shenker, and L. Zhang, "MACAW: A media access protocol for wireless LANs." In *Proceedings, 1994 SIGCOMM Conference*, London, UK.
18. M. Stojanovic, "Acoustic (underwater) communications." In J. G. Proakis, editor, *Encyclopedia of Telecommunications*. New York: John Wiley and Sons, 2003.
19. J. K. Yeo, Y. Lim, and H. H. Lee, "Modified MAC protocol design for the acoustic-based underwater data communication," *Proc. of IEEE ISIE 2001*.
20. I. F. Akyildiz, D. Pompili, and T. Melodia, "Challenges for efficient communication in underwater sensor networks," *ACM Sigbed Review*, Vol. 1, No. 2, July 2004.
21. J. H. Cui, J. Kong, M. Gerla, and S. Zhou, "Challenges: building scalable mobile underwater sensor," *IEEE Network, Special Issue on Wireless Sensor Networking*, June 2006.
22. I. Vasilescu, K. Kotay, D. Rus, M. Dunbabin, and P. Corke, "Data collection, storage and retrieval with an underwater sensor network." In *Proceedings of SenSys '05*, pp. 154–165.
23. J. Preisig, "Acoustic propagation consideration for underwater acoustic communication network development," *SenSys*, 2005.
24. I. F. Akyildiz, D. Pompili, and T. Melodia "State of the art in protocol research for underwater acoustic sensor networks." In *Proc. of ACM International Workshop on Under-Water Networks (WUWNet)*, Los Angeles, CA, September 2006.
25. D. Pompili and T. Melodia, "Three-dimensional routing in underwater acoustic sensor networks." In *Proc. of ACM International Workshop on Performance Evaluation of Wireless Ad Hoc, Sensor, and Ubiquitous Networks (PE-WASUN)*, Montreal, Canada, October 2005.
26. R. Fourer, D. M. Gay, and B. W. Kernighan, "AMPL: A Modeling Language for Mathematical Programming." Duxbury Press/Brooks/Cole Publishing Company, 2002.
27. Y. Xiao and J. Rosdahl, "Performance analysis and enhancement for the current and future IEEE 802.11 MAC protocols," *ACM SIGMOBILE Mobile Computing and Communications Review (MC2R), Special Issue on Wireless Home Networks*, Vol. 7, No. 2, Apr. 2003, pp. 6-19.
28. Y. Xiao, "MAC performance analysis and enhancement over 100 Mbps data rates for IEEE 802.11." In *Proceedings of The IEEE Vehicular Technology Conference* (IEEE VTC2003 Fall), Oct. 6-9, 2003, Orlando, Florida, USA, pp. 1869–1873.

29. Y. Xiao, "Concatenation and piggyback mechanisms for the IEEE 802.11 MAC." In *Proceedings of IEEE Wireless Communications and Networking Conference 2004* (IEEE WCNC 2004), pp. 1636–1641.

30. Y. Xiao, "Packing mechanisms for the IEEE 802.11n wireless LANs." In *Proceedings of The IEEE Global Telecommunications Conference 2004* (IEEE GLOBECOM 2004), pp. 7198–7198.

31. Y. Xiao, "IEEE 802.11 performance enhancement via concatenation and piggyback mechanisms," *IEEE Transactions on Wireless Communications*, Vol. 4, No. 5, Sept. 2005, pp. 2182–2192.

32. T. Li, Q. Ni, T. Turletti, and Y. Xiao, "Performance analysis of the IEEE 802.11e block ACK scheme in a noisy channel." In *Proceedings of IEEE International Conference on Broadband Networks* (IEEE BROADNETS 2005).

33. Q. Ni, T. Li, T. Turletti, and Y. Xiao, "Saturation throughput analysis of IEEE 802.11 wireless networks in error environments," *Journal of Wireless Communications and Mobile Computing, Special Issue on Modeling and Performance Evaluation of Radio Resource QoS for Next-Generation Wireless and Mobile Networks*, New York: John Wiley & Sons, Vol. 5, No. 8, Dec. 2005, pp. 945–956.

34. Y. Xiao, "IEEE 802.11n: enhancements for higher throughput in wireless LANs," *IEEE Wireless Communications*, Dec. 2005, pp. 82–91.

35. Y. Xiao and X. Shen, "Adaptive ACK schemes of the IEEE 802.15.3 MAC for the ultra-wideband system." In *Proceedings of IEEE Consumer Communications and Networking Conference 2006* (IEEE CCNC 2006), pp. 548–552.

36. Y. Xiao, "Efficient MAC strategies for the IEEE 802.11n wireless LANs," *Journal of Wireless Communications and Mobile Computing*, New York: John Wiley & Sons, Vol. 6, No. 4, pp. 453–466, June. 2006.

37. Y. Xiao, X. Shen, and H. Jiang", "Optimal ACK schemes of the IEEE 802.15.3 MAC for the ultra-wideband system," *IEEE Journal on Selected Areas in Communications*, Vol. 24, No. 4, April 2006, pp. 836–842.

38. T. Li, Q. Ni, D. Malone, D. Leith, Y. Xiao, and T. Turletti, "Aggregation with fragment retransmission for very high-speed WLANs," *IEEE/ACM Transactions on Networking*, Vol. 17, No. 2, 2009, pp. 591–604.

SOFTWARE, HARDWARE, AND CHANNEL MODELING

V

Chapter 10

Software-Driven Underwater Acoustic Sensor Networks

Raja Jurdak

Contents

Deployment efforts of underwater acoustic sensor networks have been impeded by the prohibitive monetary cost and high power consumption of existing acoustic hardware, which typically targets deep-water long-range communication. To address this issue, we propose software-driven underwater networks that rely on widely available speakers and microphones in electronic devices, coupled with software modems, to establish wireless acoustic communication links. The proposed acoustic communication system paves the way for cheap and easily deployable underwater acoustic sensor networks for marine monitoring applications with low data transfer requirements. This chapter presents the components that comprise the underlying communication system for software-driven underwater sensor networks, with an eye towards their application opportunities. Based on our recent experiments in rivers, canals, and ponds, we also address the technical challenges for software-driven underwater sensor networks, including modulation, symbol synchronization, filtering, and demodulation, as well as logistical challenges, such as calibration, fouling, and waterproofing, which typically arise in underwater environments. The discussion yields suitable guidelines for next steps in software-driven underwater sensor network research and development.

10.1 Introduction

Recent improvements in the processing power, memory size, form factor, and battery consumption of sensor modules have fueled increased interest in the development and deployment of underwater acoustic sensor networks. The diversity in the potential application space for underwater sensor networks, including oil prospecting, seismic and environmental monitoring, and military applications, has led to related projects with a wide range of design requirements.[3–6,21]

Interest in monitoring aquatic environments with sensor networks stems from their capability to provide sensorial data at unprecedented spatial and temporal scales. So far, aquatic monitoring platforms have attempted to capture information at high temporal scale (for example, surface buoys with suspended probes in the water, satellites that observe large geographic regions) or high spatial scale (for example, research vessels that survey sea floors). Until the recent advent of sensor networks, monitoring aquatic environments at both high temporal and spatial scales had remained an elusive goal, despite the strategic significance of water management to the social and economic development of the global population.

Most underwater acoustic sensor network efforts rely on specialized hardware for modulating, transmitting, receiving, and demodulating acoustic signals. The specialized modulation hardware ranges from powerful and expensive commercially available acoustic modems[7,8] to dedicated integrated circuits[4] and dedicated DSP (Digital Signal Processing) boards.[9–11] The communication hardware ranges from specialized underwater acoustic transducers and hydrophones[5] to generic speakers and microphones.[4] The use of specialized hardware for establishing acoustic communications underwater typically increases the network cost, the design time spent in interfacing node hardware components, and the size and weight of individual network nodes.

The focus of recent efforts on hardware acoustic modulation considers that low processing speeds do not allow the modulation of acoustic signals in software. However, modulation and demodulation of acoustic signals in software demonstrated its success in both aerial[12,14] and underwater[15,16] environments. Software modulation is an alternative approach to expensive hardware modems that overcome most of the cost and complexity drawbacks. Recent advances in miniaturization and circuit integration have yielded smaller and more powerful processors that are capable of efficiently running acoustic modulation and demodulation software. Software modulation also provides a higher level of flexibility for on-the-fly tuning of modulation parameters to suit different environments. The transmission and reception of the software modulated acoustic signal can also avoid using specialized hardware through generic speakers and microphones.

Eliminating the need for specialized hardware for acoustic communication greatly reduces the cost of network nodes, which facilitates the dense deployment of mote-class nodes to form underwater sensor networks. Within this context, this chapter proposes *software-driven underwater acoustic sensor networks* for dense shallow water quality monitoring in rivers, bays, estuaries, and lakes. The network consists of affordable off-the-shelf sensor modules (motes) that use software modems and generic hardware to communicate acoustically and send the data towards the base station through multi-hop communication. The motes are placed into elastic latex membranes that waterproof the hardware while maintaining acoustic coupling with the water channel. In addition to being cost-feasible and satisfying the temporal and spatial scale requirements, sensor motes, which were originally designed for terrestrial applications, combine processing and storage capabilities that provide an intelligent platform for network self-configuration and self-management. Finally, the speakers on board the mote platform have low output power, which is favorable for both network longevity and for minimizing interference with aquatic ecosystems.

This chapter discusses the design, application opportunities, and challenges of software-driven underwater sensor networks. Software-driven underwater sensor networks target long-term shallow aquatic monitoring applications, and they involve the design and development of the acoustic communication links, communication protocols, and application behavior. In particular, this chapter investigates the design and development of reliable acoustic communication for realizing software-driven underwater sensor networks.

The remainder of this chapter is organized as follows. Section 10.2 discusses the related work on underwater network architectures and on both hardware and software acoustic modems. Section 10.3 presents the network architecture and focuses on the physical layer functionality and higher layer communication protocol issues. Section 10.4 discusses the logistical challenges that face software-driven underwater sensor network deployments, while Section 10.5 concludes the chapter.

10.2 Related Work

This section first reviews existing architectures for underwater networks. The latter part of the section discusses acoustic modulation, both in hardware and in software, as an enabler of wireless underwater communication.

10.2.1 Underwater Network Architectures

Most underwater network proposals rely on wireless acoustic communication between a set of underwater nodes, which eventually relay information to a node at the surface. Akyildiz et al.[3] define both 2-dimensional and 3-dimensional architectures for underwater networks. The 2-dimensional architecture follows a clustered topology in which each group of underwater nodes that are fixed to the sea floor communicates with more powerful nodes, or clusterheads, in their vicinity. The clusterheads then relay the data to a surface node, which in turn forwards the data through satellite or long-range radio links to a central repository. The 3-dimensional architecture considers underwater sensor nodes that are anchored to the sea floor, thus allowing them to float at different depths depending on currents and tides. Since our chapter considers shallow water applications with communication at very short ranges, differences in depth are relatively small so the network adheres to the 2-dimensional architecture.

Another recent survey by Cui et al.[2] classifies underwater network architectures according to their deployment duration and the criticality of data they sense, differentiating between two architecture classes: long-term non-critical deployments, and short-term critical deployments. Their article also identifies very short-range underwater acoustic modems as a gap in the current state of the art. Our work coincides with the long-term non-critical monitoring architecture of Cui et al., and in this paper we specifically target the development and investigation of low-power acoustic modems for very short-range underwater networks.

Vasilescu et al.[4] propose another network architecture where autonomous underwater vehicles (AUVs) periodically visit the network area to collect data from the *in-situ* sensors through ultra short wireless acoustic links. Although our work focuses on the adaptation of stationary mote platforms for the underwater environment, the development of acoustic communication links that operate with a range of about 20 m on the motes would enhance Vasilescu's architecture, reducing the localization and navigation requirements, as well as the duration of the data collection process by the AUV.

10.2.2 Hardware Modems

Earlier efforts in acoustic communication have focused on using specialized and dedicated hardware for underwater acoustic modulation and demodulation. Acoustic underwater communication is a mature field and there are several commercially available underwater acoustic modems.[7,8] The commercially available acoustic modems provide data rates ranging from 100 bps to about 40 Kbps, and they have an operating range of up to a few kilometers and an operating depth in the order of thousands of meters. The cost of a single commercial underwater acoustic modem is at least a few thousand U.S. dollars. The prohibitive cost of commercial underwater modems has been an obstacle to the wide deployment of dense underwater networks, until the recent development of research versions of hardware acoustic modems.

Researchers at the Woods Hole Oceanographic Institution are developing a Utility Acoustic Modem (UAM) as a completely self-contained, autonomous acoustic modem capable of moderate communication rates with low power consumption.[10] This modem uses a single specialized DSP board with on-board memory and batteries. The purpose of developing the UAM is to make a more affordable acoustic modem available for the research community. Researchers at the University of California, Santa Barbara, are also developing a hardware acoustic underwater telemetry modem[9] for ecological research applications, using a DSP board with custom amplifiers, matching networks, and transducers. Their modem is intended for interfacing to nodes in an underwater ad hoc network, and it achieves a 133 bps data rate. Whereas both of the above efforts[9,10] aim at making underwater acoustic modems cheaper and more accessible by developing specialized affordable hardware, our work aims at driving the cost even lower and at making acoustic underwater communications even more pervasive through the development of software acoustic modems that can operate on generic hardware platforms.

In a more recent article, Wills et al.[11] propose their design for an inexpensive hardware modem for dense short-range underwater sensor networks. Their work aims at borrowing communication concepts, such as wake-up radio, from terrestrial sensor networks. Although we share the same end goal as Wills et al. (inexpensive acoustic modems for dense short-range wireless networks), our approach differs in its emphasis on modulation through software rather than through specialized hardware.

One of the few attempts to deal with generic microphones and speakers is by Vasilescu et al.[4] These authors propose a network that combines acoustic and optical communications, stationary nodes, and AUVs for monitoring coral reefs and fisheries with ranges in the order of hundreds of meters. Their work uses generic microphones and speakers along with a specialized integrated circuit that generates amplitude shift keying (ASK) or frequency shift keying (FSK) modulated sound signal in order to demonstrate the acoustic communication capability under water.

Vasilescu et al. achieve a bit rate in the order of tens of bits per second up to about 10 to 15 meters. Although our work resembles their work in the use of generic microphones and speakers for acoustic communications, it differs in its proposal and implementation of software modems for off-the-shelf mote platforms rather than the use of specialized integrated circuits for communication.

10.2.3 Software Modems

With the rapid increase in processor speeds, the idea of implementing acoustic modems in software becomes feasible and even attractive due to the low cost processing power. Coupling software acoustic modems with the use of microphones and speakers for transmission and reception can eliminate the need for specialized hardware for acoustic communication, trading off cheap computational power for expensive communication hardware. The cost of software acoustic modems is limited to the development cost, after which the per-unit cost is zero.

Because of these attractive features, Lopes and Aguiar[12] have investigated using software modems for aerial acoustic communications in ubiquitous computing applications. Building on their work, software acoustic modems can also eliminate the need for specialized hardware in underwater acoustic communications, thereby encouraging wider deployment of underwater sensor networks. In preliminary experiments, we started profiling the underwater acoustic spectrum and data communications capabilities with software acoustic modems.[15] Our work used waterproofed generic microphones and speakers, connected to laptops on the surface, for sending and receiving software modulated acoustic signals. The achieved bit rates were in the order of tens of bits per second for distances up to 10 meters. Our work here extends our earlier work by coupling software modems with Tmote Invent module hardware. Our recent study[16] presents underwater experiment results that confirm the communication capability of software modems on Tmote Invent hardware. This chapter further investigates the application opportunities and the technical and logistical challenges of this communication system.

10.3 Software-Driven Underwater Acoustic Sensor Networks

This section describes the architecture for software-driven underwater sensor networks. Figure 10.1 sheds more light on the target network application.

10.3.1 Network Overview

Software-driven underwater sensor networks will consist of tens to hundreds of motes deployed in a shallow water environment. The motes can communicate acoustically through short-range wireless links, thanks to their on-board speaker

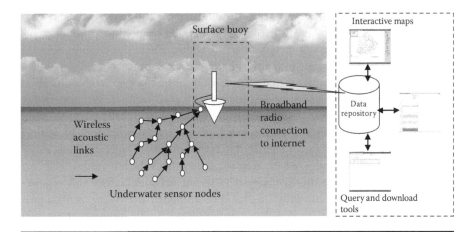

Figure 10.1   Target network application.

and microphone and to software acoustic modems. The motes are placed into elastic latex membranes, which maintain the acoustic coupling of the on-board speaker and microphone with the water, while waterproofing the motes. The modules periodically sample their sensors, collecting physical indicator data from their built-in sensors, such as water temperature or tidal strength, or from add-on sensors, such as salinity or phosphorous levels, which help determine water contamination levels. After sampling their sensors, the nodes send the sensor values to neighboring nodes, which in turn relay the data through multi-hop links to the nearest collection point. Because each node periodically sends only a few 2-byte sensor values, *the data transfer rate requirements of the network are low.*

The network architecture supports two embodiments of the collection point: (1) an underwater instrumentation station that relays network data to a data repository on shore through a sub-sea fiber optic cable, or (2) a surface station that resides on a buoy and relays network data to the data repository on shore through a long-range RF communication link. For both configurations, the logical network topology is a tree, or more generally, a multiple-tree topology, in the case of several collection points.

10.3.2 Motivating Application

Software-driven underwater acoustic sensor networks support shallow water aquatic application scenarios that require long-term dense monitoring of a given physical area.

A representative scenario is the monitoring of inland water bodies, such as rivers, lakes, and estuaries. In particular, IBM and the Beacon Institute have recently announced a joint project to monitor the Hudson River and to provide real-time minute-by-minute environmental data from within the river.[1] The collected environmental data includes salinity, temperature, dissolved oxygen, and pollution loading. This project aims to develop multiple new technologies and mechanisms for capturing, collecting, aggregating, disseminating, and analyzing data. One challenge in this project is the dimensions of the area to be monitored, which spans a long distance with a more or less constant width. These dimensions, coupled with numerous natural bends in the river path, preclude the use of traditional oceanic communication paradigms that assume an open line-of-sight communication channel. Short-range communication systems, such as software-driven underwater sensor networks, provide an appropriate monitoring solution for river monitoring, as nodes can use multi-hop communication links and bypass obstacles such as river bends. Furthermore, the use of short-range communication links demands dense deployments of sensor nodes, which is cost-feasible only if the individual node cost is small. Again, software-driven underwater sensor networks satisfy this requirement, as the unit cost is currently small and continuously decreasing as mass production of sensor nodes rises. The dense deployment actually improves the spatial granularity of the river monitoring application, because it increases the available data points for the same physical area.

10.3.3 Target Platform

For our application, we have selected mote-class computers, which are powerful enough to perform sufficient in-network processing and are affordable enough to enable the deployment of a dense network at reasonable cost. In particular, we have selected the Tmote Invent module (shown in Figure 10.2), which has an onboard SSM2167 microphone from Analog Devices that is sensitive to frequencies

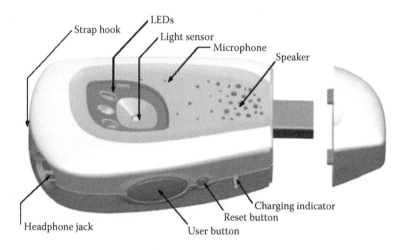

Figure 10.2 The Tmote Invent module. (From Jurdak et al. Design Considerations for Deploying Underwater Sensor Networks, *Proceedings of IEEE Conference on Sensor Communication, *SENSORCOMM.* UNWAT Workshop, Valencia, Spain, October. © 2007 IEEE. With permission.)

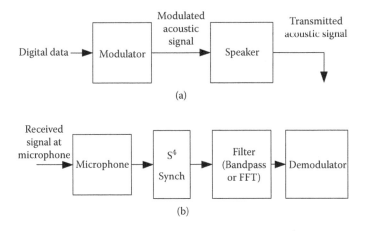

Figure 10.3   Block diagram for software modem. (a) Modulator/Transmitter. (b) Demodulator/Receiver. (From Jurdak et al. Design Considerations for Deploying Underwater Sensor Networks, *Proceedings of IEEE Conference on Sensor Communication, *SENSORCOMM.* **UNWAT Workshop, Valencia, Spain, October. © 2007 IEEE. With permission.)**

from 100 Hz to 20 kHz, and an on-board TPA0233 speaker amplifier from TI with an 8-ohm speaker that has a range of 400 Hz to 20 kHz. The goal of this work is to exploit the on-board microphone and speaker to establish short-range acoustic links among Invent modules.

10.3.4 System Components

Figure 10.3 illustrates the main components of the communication system. At the sender side, digital sensor data from the underwater environment first enter the software modulator, resulting in a modulated acoustic signal. The on-board speaker then transmits this signal into the underwater channel. At the receiver, the on-board microphone captures the signal and the resident software performs symbol synchronization through the S^4 block,[17] filtering through Fast Fourier Transform (FFT) or wavelet decomposition, and finally demodulation.

10.3.4.1 Modulation

The first component of the acoustic communication system is software modulation, which takes digital data as input and modulates an acoustic signal with the data. The potential choices of modulation schemes for software modems include ASK, phase shift keying (PSK), and FSK. We have selected FSK as the lowest complexity mechanism to run on the resource-limited mote platforms.

In general, FSK uses 2^N frequencies to encode N bits per frequency. The signal demodulation at the receiver can use low complexity techniques, such as the FFT or wavelet decomposition, to determine the frequency content of the signal. Choosing a number of frequencies that have high signal-to-noise ratio (SNR) for the channel and ensuring sufficient spectral separation between the frequencies for FSK provides robust underwater communication for low power transmissions with minimal processing complexity at the receiver.

We recently performed an empirical study to investigate the spectral properties of the underwater channel in a controlled water environment.[16] The study used the Tmote Invent[18] speaker as the transmitter of acoustic signals and a generic PC microphone as the receiver. The components were waterproofed using off-the-shelf elastic latex membranes that vibrate sufficiently to preserve most the acoustic properties of the speaker and microphone. The study evaluated the SNR of frequency tones between 400 and 6700 Hz at 100-Hz increments. The selection of the 100-Hz band separation between frequencies provides for low complexity frequency detection at the receiver. Note that narrower separation bands enable the use of more frequencies within the same available bandwidth, which increases bit rates but also reduces signal quality at the receiver. In general, there is a tradeoff in digital modulation techniques between the number of signal levels (in phase, amplitude, or frequency) and the quality of the signal. In our case, increasing the number of frequencies by using narrower frequency bands requires the use of higher-order filters or FFT at the mote to decode the signal, which is a processing-intensive task. This is not favorable to the limited processing power of the motes.

The study revealed that the channel, which includes the speaker, latex membranes, the water, and the microphone, exhibits the highest SNR at frequencies in the range of the 1000 Hz to 2500 Hz. The SNR drops steadily at frequencies above 3 kHz.

The transmitted signal is received by the microphone at a remote node, on which the software then proceeds with the decoding process. Essential to acoustic communication with software modems is the ability of the receiver to synchronize to the first symbol of an incoming data stream. Traditional symbol synchronization approaches rely on the transmission of a predefined sequence of symbols, often referred to as a training sequence. The conventional approach makes two assumptions about the communication channel that do not hold for software-driven acoustic communications, namely: (1) a bit rate at least in the order of tens of kilobits per second; (2) a bit error rate (BER) in the order of 10^{-6} or lower. Software-driven acoustic communication, both aerial and underwater, supports lower bit rates that range between tens to hundreds of bits per second.[15,16] In addition, the BER of software-driven acoustic communication is several orders of magnitude higher than the radio frequency BER. The higher BER in acoustic communications tends to cause loss of training sequence symbols, preventing proper symbol synchronization. Furthermore, providing high redundancy in the training sequence to

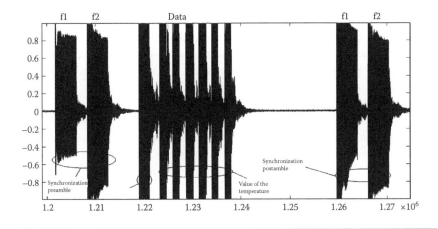

Figure 10.4 A sample showing the signal shaper as generated with the Tmote Speaker, with the preamble and post-amble S^4 symbols at the beginning and at the end of the signal.

mitigate training symbol losses is not an option for the narrow usable bandwidth of software-driven acoustic communications.

We recently proposed and tested a synchronization mechanism[17] that uses Short Signature Synchronization Symbols (S^4) to align symbol boundaries at the receiver. The design of the signature symbol aims at a high probability of correlation at the receiver even in cases of partial loss of the symbol and at low probability of false synchronization with ambient noise or data symbols. To this end, the symbol features include two square waves, with each square wave transmitted at a predefined frequency, separated by a predetermined guard time. Figure 10.4 shows a sample signal with an S^4 preamble and post-amble. The use of two frequencies for the signature symbol mitigates the effects of frequency selective fading or interference. The signature symbol guard time duration is chosen so that it is not equal to, and not a constant multiple of, the inter-symbol guard times to avoid high correlation with data symbols. The output of the S^4 provides the index of the first data sample in the received signal.

10.3.4.2 Filtering

The receiver then proceeds to filter the received signal. Filtering the acoustic signal at the receiver minimizes the effect of out-of-band noise on the decoding process. A suitable choice of filtering method depends highly on the processing capability of the receiver. Our system currently provides two filtering methods, with the first method employing narrow bandpass filters at the relevant frequencies. The second filtering method applies an FFT to the received signal and examines the signal amplitude at the FFT samples corresponding to the relevant frequencies.

The narrowband filtering method provides a finer signal quality as it only focuses on the frequencies of interest and excludes all interference outside this spectrum. The superior performance of the narrowband filtering method comes at the cost of higher processing activity at the receiver. In fact, the narrowband filtering method is suitable for running on a PC or a PDA that acts as a base station for the underwater network, as in Figure 10.1.

The FFT method can run on the motes themselves as it has lower processing complexity. Running the FFT method on motes enables the deployment of a multi-hop network where nodes can autonomously decode, process, and relay received signals. The FFT method provides a coarser signal quality than filtering since it does not exclude interference from outside the frequency spectrum of interest.

10.3.4.3 Demodulation

The filtered signal then proceeds to the demodulator component. The demodulator begins examining the signal at the first data sample, which has been determined by the S^4 block. Starting at the first data sample and taking the number of samples that corresponds to one symbol, the demodulator determines the frequency component with the highest amplitude within this window, and outputs the data symbol corresponding to the highest frequency. For subsequent bits, the demodulator shifts the start sample of the previous symbol by the symbol length, and repeats the process of determining the strongest frequency component.

10.3.4.4 Communication Protocols

The design of the higher layer communication protocols for this system should adopt a minimalist approach to avoid creating high overhead in this bandwidth- and energy limited system. The medium access control (MAC) protocol design for underwater sensor networks cannot exploit traditional low duty cycle protocols for terrestrial sensor networks. The reason is that the energy cost of transmission is generally much higher than the cost of signal reception in underwater networks, whereas the cost of transmission and reception is almost the same in terrestrial sensor networks.[19] Thus, signal transmissions dominate the energy consumption profile of underwater nodes. As such, the MAC protocol design should minimize control overhead messages, such as request to send (RTS) and clear to send (CTS), rather than implementing sleep policies. One possibility is to use burst tones at the beginning of transmission to reserve the channel. In particular, the S^4 preamble can serve as a burst tone for reserving the channel at the MAC layer.

For routing data towards the surface node, our communication system advocates the use of short multi-hop links for deploying dense underwater sensor networks. Keeping in line with the system's minimalist approach, the network should make use of simple multi-hop routing techniques, such as the directed broadcast with overhearing method of MERLIN.[20] The relatively low cost of receiving signals actually favors the use of overhearing. Our recent study on the upper bounds for transmissions in directed broadcast networks with overhearing has shown that this method has only a few redundant packets, resulting in low overhead for large networks.

The design of all communication protocols for software-driven underwater sensor networks must follow a minimalist cross-layer[22] approach that couples layer functionalities when possible in order to efficiently utilize the scarce bandwidth and energy resources of the nodes. For example, the synchronization symbols in S^4 may serve the dual purpose of node identification, typically a MAC layer feature, and symbol synchronization, a physical layer mechanism. Such cross-layer optimization reduces the control overhead of the communication system, enabling better utilization of the available channel bandwidth.

10.4 Logistical Considerations

This section discusses the logistical challenges that face software-driven underwater sensor network deployments.

10.4.1 Waterproofing and Casing

The most common waterproofing method for underwater communications hardware is to place the hardware in a custom-designed waterproof case with a special air-locked hole for the hydrophone and transducer that are in contact with the water. The cost associated with the material and design of the custom-designed casing strategies significantly increases system cost and discourages large-scale deployments of underwater sensor networks.

Our project's design strategy advocates the use of off-the-shelf components not just for communication and sensing, but also for the protection and waterproofing of the components. As such, our design places the sensor nodes in elastic latex membranes that are cheap and readily available on the market. The latex membranes take the form of a balloon that is sealed at the mouth to waterproof the sensor nodes. Since the sensor module includes the speaker and microphone, these acoustic communications components are also fully enclosed within the latex membranes. The elasticity of the membranes ensures that the acoustic waves transmitted by the speaker are transferred to the water through the elastic membrane. At the receiver side, the membranes vibrate upon the reception of an acoustic signal, transferring the signal from the water channel to the interior of the membrane where the microphone can detect it.

The use of the latex membranes causes relatively small reductions in signal amplitude. Our recent study compared the suitability of two membranes for waterproofing the sensor nodes: (1) a vinyl membrane, and (2) a latex membrane.

Table 10.1 Percentage of Correctly Received Symbols

Transfer Rate (bps)	12	24	48
Latex mem. up to 17 m	≥ 90%	≥ 81%	≥ 79%
Vinyl mem. up to 10 m	≥ 90%	≥ 78%	≥ 35%

Source: Jurdak et al. Design Considerations for Deploying Underwater
Sensor Networks, *Proceedings of IEEE Conference on Sensor
Communication, *SENSORCOMM.* UNWAT Workshop, Valencia,
Spain, October. © 2007 IEEE. With permission.

Table 10.1 illustrates the percentage of correctly received symbols for bit rates ranging between 12 bps and 48 bps for both membranes. The vinyl membrane experiments were conducted with a generic PC speaker as a transmitter, while the latex
membrane experiments were conducted with Tmote Invent speakers. The output
power rating of the two speaker types is the same, providing solid ground for comparing the results of the two experiment sets. The results in Table 10.1 show that
the latex membrane has a better coupling with the water, yielding a notably lower
BER at the higher transmission rate of 48 bps.

10.4.2 Calibration

Our field tests in different bodies of water have revealed a distinct background noise
and interference pattern in each case. For instance, the primary noise source in swimming pools is water pumps, whereas the noise sources in a river include downstream
currents and wildlife activity. The dependence of the noise profile on the deployment
environment requires calibration steps, which could be manual or automatic, prior to
placing the sensors in the water. Fortunately, the implementation of modulation and
communication in software provides maximum flexibility for on-the-fly calibration.

A central issue for calibration is the frequency-selective noise in the deployment
environment. The choice of frequencies for the S^4 synchronization symbol must
avoid frequencies with high noise in a particular deployment environment. The
selection of the S^4 frequencies is critical for proper system operation, as choosing
unsuitable frequencies causes large synchronization errors, resulting in many bit
misalignments. Similarly, it is also important to choose data symbol frequencies
that avoid the high noise frequencies.

The selection of the suitable frequencies for S^4 synchronization and the data
symbols can be done automatically. During an initial setup phase, one node, typically the base station, is designated as a calibration receiver, and another as a calibration transmitter. Upon deployment, the designated transmitter sends an *a priori*
known calibration signal that includes a diverse set of S^4 symbols with different frequency combinations, followed by a sequence of frequency tones that covers all the
possible symbol frequency tones. The designated receiver processes the calibration

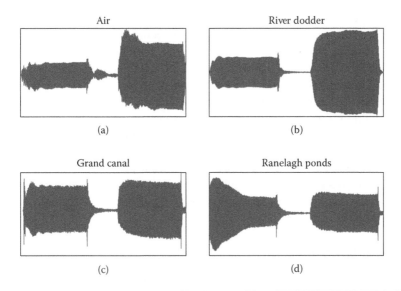

Figure 10.5 **Shape of S^4 symbol as received in (a) indoor aerial channel; (b) River Dodder; (c) Grand Canal in Dublin; (d) Ranelagh pond.**

signal by comparing the processed signal against a locally stored version of the reference signal. The receiver then selects the frequencies that have been received with the highest SNR, and transmits a short broadcast message indicating these frequencies to the other nodes.

Another calibration issue is ensuring that the stored reference S^4 symbol at each node is representative of the symbol as it is received in the current deployment environment. Our field experiments have shown that the structure of a received S^4 symbol with the Tmote Invent microphone in air is different than its structure in water. As Figure 10.5 shows, the structure even varies across different water bodies, depending on the current, depth, and suspended solids in the water. For instance, the envelope of the S^4 symbol differs significantly when the signal is received in the river (Figure 10.5b) and in the pond (Figure 10.5d). The plots in Figure 10.5 illustrate that the relative amplitude of each of the square signals changes depending on the medium. Other signal artifacts, such as impulses at the beginning or at the end of the signal, are also dependent on the deployment environment. As such, each node should store at least one instance of the S^4 symbol *as it is received in the current deployment environment.* This maximizes the probability of successful correlation and synchronization through the S^4 mechanism.

10.4.3 Fouling

Fouling is a process by which marine wildlife, such as barnacles, zebra mussels, weeds, and algae, attach themselves to a still object in the water. Avoiding fouling effects is especially important for software-driven underwater sensor networks since the attachment of organisms to the membrane could limit or change the membrane's vibration characteristics. Traditional anti-fouling approaches include the use of special copper-based paints to prevent the attachment of organisms to boat bottoms. Current research focuses on developing alternatives[23] to paint-based solutions, which are harmful to the ambient environment.

Our project aims at protecting the environment, and not damaging it in the process, so we intend to adopt one of the emerging anti-fouling techniques. One interim solution under consideration is to place the nodes in the latex membranes and then to fix the membranes inside a resilient cubic plastic box whose purpose is to shield the nodes from fouling and other hazards in harsh underwater environments. The surfaces of the plastic box would be perforated to maintain acoustic coupling between the box contents and the water.

10.4.4 Deployment

Software-driven underwater sensor networks target long-term monitoring applications, which demand suitable strategies for deployment, tracking network health, and performing network maintenance when necessary. First, the deployment strategy must ensure that the nodes are well anchored in the environment, so that tides, waves, and marine wildlife do not cause the nodes to move and lose connectivity with neighboring nodes. Anchoring methods are highly dependent on the deployment environment. In a river, nodes may be anchored either to the river floor through attachment to a heavy object, or to the river bank through suspension with a solid pole that is securely fixed to the river bank. In larger bodies of water, the former anchoring method is preferable since most points of interest will be far from the water edges.

In addition to anchoring, deciding where to deploy sensor nodes is also highly dependent on the application scenario. In the river monitoring application, sensors can be placed at regular distance intervals all along the river. When monitoring water quality, sensor placement should take into account potential pollution sources in the river, such as pipes or construction sites that may dump pollutants into the water. In some instances, providing redundant sensors immediately upstream and downstream from the pollutant sources can prove crucial for delivering timely alerts to water management officials on pollution events, such as the dumping of waste material into the river.

10.5 Conclusion

This chapter introduced software-driven underwater acoustic sensor networks as an enabler of dense shallow water wireless network deployments. Building on our practical experience with underwater communication experiments in the River Dodder in Dublin, the discussion focused on river monitoring as a motivating application scenario for software-driven underwater sensor networks. The system components of software-driven acoustic networks were then presented, including the signal modulation, synchronization, filtering, and demodulation techniques for software modems. Looking ahead towards a full network stack based on software modems, the design of higher layer communication protocols for this system should also adopt a cross-layer design approach to minimize control overhead. Logistical considerations for underwater networks depend on the deployment environment. We have identified common logistical considerations for low-power underwater networks, which include the need for resilient waterproofing and casing, calibration, deployment strategies, and anti-fouling measures. Our initial experiments in various bodies of water have exposed the benefits of using software modems for underwater communications capable of functioning on off-the-shelf multi-purpose sensor modules. The design guidelines in this chapter lay the groundwork for further development of software-driven underwater sensor networks.

References

1. B. Sauser. "Networking the Hudson River," *Technology Review*, Published by MIT. August, 2007. Available at: http://www.technologyreview.com/Infotech/19309/
2. J. Cui, J. Kong, M. Gerla, and S. Zhou. "Challenges: Building Scalable Mobile Underwater Wireless Sensor Networks for Aquatic Applications," *IEEE Network*, 20(3):12-18, 2006.
3. I. F. Akyildiz, D. Pompili, and T. Melodia. "Underwater Acoustic Sensor Networks: Research Challenges," *Ad Hoc Networks* (Elsevier), vol. 3, no. 3, pp. 257-279, March 2005.
4. I. Vasilescu, K. Kotay, D. Rus, M. Dunbabin, and P. Corke. "Data Collection, Storage, and Retrieval with an Underwater Sensor Network." In *Proc. Sensys '05*, San Diego, CA, 2005.
5. J. Heidemann, Y. Li, A. Syed, J. Wills, and W. Ye. "Underwater Sensor Networking: Research Challenges and Potential Applications," USC/ISI Tech. Rep. ISI-TR-2005-603, 2005.
6. X. Yang et al. "Design of a Wireless Sensor Network for Longterm, In-Situ Monitoring of an Aqueous Environment," *Sensors*, 2:455-472, 2002.
7. Linkquest Inc. Available at: www.link-quest.com
8. DSPComm. Available at: www.dspcomm.com
9. R. A. Iltis, H. Lee et al. "An Underwater Acoustic Telemetry Modem for Eco-Sensing." In *Proc. Oceans '05*, September 2005.
10. Utility Acoustic Modem. Available at: auvlab.mit.edu
11. J. Wills, W. Ye, and J. Heidemann. "Low Power Acoustic Modem for Dense Underwater Sensor Networks." In *Proc. IEEE (WUWNet06)*, 2006.
12. C. V. Lopes and P. Aguiar. "Acoustic Modems for Ubiquitous Computing," *IEEE Pervasive Computing*, 2003.
13. Blauert, Jens (Ed.). *Communication Acoustics*. Berlin: Springer, 2005.
14. R. Jurdak, P. M. Q. Aguiar, C.V. Lopes, and P. Baldi. "A Comparative Analysis and Experimental Study on Wireless Aerial and Underwater Acoustic Communications." In *Proceedings of the IEEE International Conference on Digital Telecommunications (ICDT)*, Cap Esterel, France. August, 2006.

15. R. Jurdak, C.V. Lopes, and P. Baldi. "Software Acoustic Modems for Short Range Mote-Based Underwater Sensor Networks." In *Proc. of IEEE Oceans,* Singapore. May, 2006.
16. R. Jurdak, P.M.Q. Aguiar, P. Baldi, and C.V. Lopes. "Software Modems for Underwater Sensor Networks." In *Proc. of Oceans '07.* June, 2007.
17. R. Jurdak, A.G. Ruzzelli, G.M.P. O'Hare, and C.V. Lopes. "Reliable Symbol Synchronization in Software-Driven Acoustic Networks." In *Proc. of IEEE GlobeCom,* Washington, D.C. November, 2007.
18. Sentilla Corporation. Available at: www.sentilla.com
19. P. Harris, M. Stojanovic, and M. Zorzi. "When Underwater Acoustic Nodes Should Sleep with One Eye Open: Idle-Time Power Management in Underwater Sensor Networks." In *Proc. WUWNet.* 2006.
20. A.G. Ruzzelli, G.M.P O'Hare, and R. Jurdak, "Merlin: Cross-Layer Integration MAC and Routing for Low Duty-Cycle Sensor Networks." To appear in *Elsevier Ad Hoc Networks Journal, Special Issue on Energy Efficient Design in Wireless Ad Hoc and Sensor Networks,* 2008.
21. R. Black. "Research Needed on Marine Sound," BBC News article. Available at: news. bbc.co.uk/2/hi/science/nature/4706670.stm
22. R. Jurdak. *Wireless Ad Hoc and Sensor Networks: A Cross-Layer Design Perspective.* Berlin: Springer-Verlag, 2007.
23. The Maya2 Project. Available at: http://www.maya-net.org/
24. R. Jurdak et al. Design Considerations for Deploying Underwater Sensor Networks, *Proceedings of IEEE Conference on Sensor Communication, *SENSORCOMM.* UNWAT Workshop, Valencia, Spain, October.

Chapter 11

HW/SW Co-Design of a Low-Cost Underwater Sensor Node with Intelligent, Secure Acoustic Communication Capabilities

Fei Hu, Yang Xiao, Paul Tilghman, Steven Mokey, James Byron, and Andrew Sackett

Contents

Underwater Wireless Sensor Networks have important applications in ocean exploration and lake pollution monitoring. They use acoustic media to achieve sensor communications. They are essentially different from terrestrial radio frequency (RF)-based sensor networks due to their highly variable, long acoustic delay, and mobility nature. This chapter describes our hardware/software co-design of low-cost underwater sensor nodes for deployment in a shallow underwater environment. Each sensor is comprised of a Digital Signal Processing (DSP) board responsible for modulation/demodulation and sensor readings, as well as an analog board responsible for signal strength amplification and signal conditioning. In this particular application the system is used to measure pH and temperature of the water environment. In addition, we have implemented an intelligent underwater ad hoc networking scheme with low-complex, secure data transmission protocols. The security scheme is based on symmetric cryptography without causing much encryption/decryption overhead.

11.1 Introduction

11.1.1 Terrestrial and Underwater Sensor Networks

Recent advances in sensor network developments promise a burgeoning new world in which large groups of sensors cooperating together provide a reliable stream of information from remote places. Sensor networks have many potential applications, ranging from habitat monitoring to military applications, such as battlefield command control and monitoring, to biomedical applications. While the field of over-the-air sensor networks has seen great growth in recent years, the development of underwater sensor networks (USNs) has remained relatively untapped. A USN deployed in lakes can be used to monitor the water quality levels, including the pH, toxicity level, salinity, etc. Additionally, there is a growing area of research using USNs for anti-submarine warfare, torpedo defense, mine countermeasures, and other underwater defense applications [1].

Terrestrial and *underwater* sensor networks, however, have some common features as follows:

1. *Low-energy consumption*: Due to their remote operation, sensor networks must be battery powered. Efficient use of this resource is crucial to maintaining a long operating lifetime of the network.
2. *Low cost*: Development of sensor networks should strive to make manufacturing costs as minimal as possible. A large-scale sensor network can be deployed over a larger area to monitor a larger number of events, or to observe the correlation between related events.

While these two types of networks share a great deal in common, there is a vast disparity, not only between their operating environments, but also between their technical developments. For example, terrestrial sensor networks use RF as wireless medium. However, RF propagation has very limited communication range (for instance, Mica-2 transmit range has been measured as less than one meter in fresh water [2]). In fact, even though long-wavelength RF can penetrate water for a longer distance, it requires large transmit power and a large antenna, making it inappropriate for small, low-power sensor nodes. Hence, underwater *acoustic* communication provides an important alternative. Moreover, acoustic sound travels faster and longer in water than in air (see [2] for a good overview on underwater acoustic communications).

Recent research into the development of energy-efficient medium access control (MAC) [3] mechanisms as well as routing protocols has been in good progress in *terrestrial* sensor network platforms, which consist of a large amount of commercial sensors such as RENE and Crossbow motes. While these systems are available for a reasonable low cost (approximately US$100) [2], commercially available underwater communication systems have an extremely high cost (US$10,000 or more) [2,4]. The necessity for a *cost-effective* underwater network is paramount.

Our developed USN to be reported in this chapter is a prototype that can provide a low-cost test-bed for short- to medium-range sensor networks. Figure 11.1 shows our targeted USN application scenario. Underwater sensors use acoustic communications to form a hop-to-hop data relay architecture. Each sensor has Analog-to-Digital (A/D), DSP, and acoustic communication modules. The underwater node closest to the surface can forward all the sensed data (collected from itself or neighboring sensors) to the surface node.

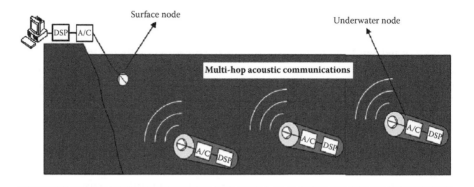

Figure 11.1 USN application scenario.

11.1.2 Underwater Communication Challenges

As mentioned before, the physical properties of water prevent RF-based wireless communication. While *light* may represent a high-bandwidth data connection, it refracts quickly with distance and does not represent a viable solution to general underwater communications applications. Additionally, light is a directional communication medium, and would not perform in an environment without situational awareness (this means that omni-directional communications are necessary). In light of the crippling deficiencies of typical transmission media, sound is used in lieu of light and RF.

Underwater acoustic communications have a variety of physical properties that make this type of communication more difficult than traditional RF. The primary challenge is the speed at which sound waves propagate under water. It is typically around 1500 m/s. RF propagates at approximately 3×10^8 m/s, five orders of magnitude faster than sound. Also, the speed of sound changes dramatically (up to 100 m/s) with changes in pressure, salinity, and temperature [2]. In shallow water, reflections off the water's surface as well as the bottom are common. This reflection can potentially continue to reverberate inside the acoustic channel, and arrive at the receiver again. This information can combine with another symbol, causing an effect known as Inter Symbol Interference (ISI) [5]. Because of the long propagation delay, one symbol may interfere with other symbols as it continues to reverberate throughout the channel. Additionally, the bandwidth of the underwater channel is effectively limited to approximately 100 kHz due to the absorptive factor of water. The frequencies higher than 100 kHz can severely limit the propagation range of the signals since higher frequencies attenuate more quickly under water.

11.1.3 Related Work

Most commercially available underwater communication systems [4,6] are designed for long-range communications (with link distances of several kilometers). These modems can carry sustained data rates of approximately 1 k ~ 40 kbps. Also, these units cost more than US$10,000. Implementation of a system is thus cost prohibitive. Additionally these systems function only as modems and they are not capable of running any of the OSI (Open System Interconnection) layers above the data link layer without the addition of another processing unit (such as a laptop or microcontroller).

In [7] the authors have described the design of a low-cost sensor node for adhoc underwater communications with an interface to a surface buoy. This system uses Amplitude Shift Keying (ASK) modulation. The authors chose to rely upon analog hardware to convert the modulated signal back to the digital representation, and vice versa. The use of ASK modulation under water is problematic due to ISI caused by reflections in the underwater channel. A special case of ASK was used called On-Off Keying (OOK) where a digital "1" is represented by the presence of a carrier wave (with certain amplitude) and a digital "0" is represented by the lack of a carrier wave. The ISI induced by the channel could easily lead to bit errors due to reflection of the carrier wave over a period when a zero should be present. Also, the system does not allow easy expansion of its current capacity and throughput without a complete redesign of the hardware.

The Woods Hole Oceanographic Institution's (WHOI's) Micro-Modem is a small underwater acoustic modem capable of transmitting at 5 kpbs [8]. The Micro-Modem has a number of transmission modes including both frequency-hopping Frequency Shift Keying (FSK) and a high rate Phase Shift Keying (PSK). The modem relies on a Texas Instruments (TI) floating point DSP to handle all aspects of modulation and demodulation. However, this modem is designed for applications with link ranges of approximately 2–4 km, and has a typical transmission power of 50 W. Its power consumption is low considering the link ranges it must achieve; however, it remains unsuitable for applications with link ranges of approximately 100 m.

The authors of [9] have created a relatively cheap underwater sensor node with software-defined FSK communication schemes using a typical terrestrial sensor node. In this application a T-mote RF sensor node was outfitted with a microphone and speaker to facilitate acoustic communications. This setup allows for the exploitation of commercially available sensors for the creation of a relatively inexpensive underwater network testbed. While capable of transmitting and receiving up to 96 bps, reliable communications (Bit Error Rate less than 10%) were achieved with a transmission speed of 12–24 bps.

With these works in mind, our system architecture (to be discussed below) utilizes a single fixed-point DSP to handle all signal processing and communications, coupled with an analog board responsible for the signal amplification and signal conditioning.

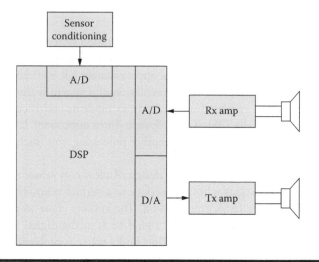

Figure 11.2   Hardware and DSP interaction.

11.2 System Architecture—HW/SW Co-Design

Our design is to provide a cost-effective development platform for underwater networking schemes. To facilitate expandability for future development, the design of the underwater sensor node relies heavily upon the separation of hardware and software.

The majority of the sensor node's functionality is software defined, allowing for easy reconfiguration of the platform for a variety of tasks. To allow for such a software-defined platform, the hardware must remain relatively simple and provide only the following functionality: amplify both outgoing and incoming signals, and provide signal conditioning for all environmental sensors and probes. This leaves the remainder of the node's functionality to be defined in software, including both modulation and demodulation.

While a variety of embedded processors could fulfill the networking and routing needs, a DSP is needed for this platform since modulation and demodulation will take place in software. Figure 11.2 shows the connection of DSP chip with other components such as Transmitter (Tx) and Receiver (Rx).

11.2.1 Hardware Design

11.2.1.1 Hydrophones

While our design focus lay in transferring traditionally hardware-centric tasks (such as modulation and demodulation) into more easily configurable software routines in order to create an end unit with relatively low cost, one area of hardware still

Figure 11.3   Two waterproofed loudspeakers serving as hydrophones.

represents a significant cost for any underwater communication platform—transducers. Hydrophones (used to convert the electrical signals into sound waves and vice versa) cost approximately US$1,000, even for small transducers designed for shorter-range applications [10]. Therefore, a cheaper alternative than commercially available hydrophones was needed. Two concepts were used to construct cheap but effective transducers.

The first concept we came up with was to utilize the back of a sports watch, with two leads soldered onto the piezzo-disc used to make the watch beep. While this alternative proved to be effective, optimal frequency response of the sports watches under test did not occur until at least 10 kHz or higher. For this application good frequency response would be needed in the 3 ~ 4 kHz range, in order to reduce the number of baseband samples taken at the analog-to-digital converter (ADC) without re-sampling. These hydrophone alternatives can be created fairly inexpensively, depending on the cost of the watch.

The second concept we used was a small loud speaker in which the paper cone was waterproofed. These speakers produced an audibly louder tone at the desired frequency range, and were used as both the transmitting and receiving hydrophones (Figure 11.3). The cost of this solution is around US$2, and allows effective communication for the desired range without the expense of traditional transducers.

11.2.1.2 Sensor Signal Conditioning

The sensors change their voltage in reaction to changes in the phenomenon they monitor. For this application two sensors were used: pH and temperature. The signal must be conditioned so that it produces a valid signal for the ADC. The signal conditioning circuits for the two sensors used are provided here.

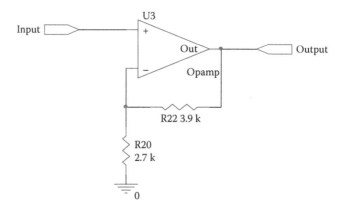

Figure 11.4 pH signal conditioning circuit.

1. *pH Signal Conditioning*: The pH amplifier has a constant gain of 2.4. The pH amplifier design consists of a non-inverting amplifier (see Figure 11.4). An operational amplifier that can accept high impedance sources was needed because the pH sensor used has an impedance of 50 MΩ, and the underwater node's ADC can only accept at most 5 kΩ sources.

2. *Temperature Signal Conditioning*: The temperature amplifier has a constant gain of 2, which covers the analog range of the ADC, and also provides greater precision. The temperature amplifier utilizes a Wheatstone bridge and an inverting differential amplifier (see Figure 11.5). The temperature sensor used is a 10 kΩ thermistor.

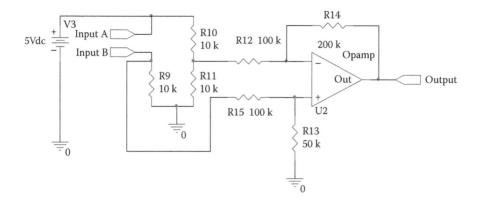

Figure 11.5 Temperature signal conditioning circuit.

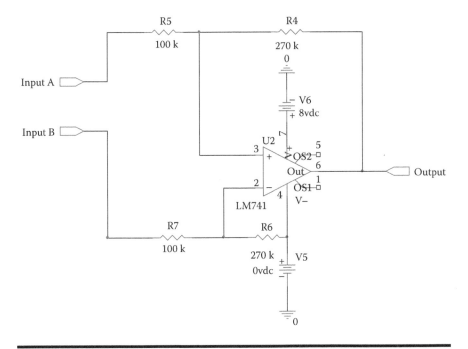

Figure 11.6 Transmitter amplifier.

11.2.1.3 Transmitter Amplifier

The transmit amplifier accepts output directly from the DSP digital-to-analog converter (DAC) interface to reach longer distances. This amplifier is composed of a summing amplifier with a 4 V DC offset (see Figure 11.6). A gain factor of 2 is used, as the expected voltage from the DAC is from –2.3 V to +2.3 V.

11.2.1.4 Receiver Amplifier

The receiver amplifier (see Figure 11.7) was designed with high gain to help reach longer distances. The gain of the receiver amplifier is variable from 60 dB to 100 dB gain. The gain is adjusted in the second stage by way of a digital potentiometer. The potentiometer can be controlled by digital control signals sent from the DSP's digital I/O lines. In this fashion the DSP can control (through software) the gain of the receiver amplifier. The amplifier has a center frequency of 3 kHz and a 3 dB bandwidth of 3 kHz.

11.2.1.5 A/D and D/A Sampling

In order to ensure appropriate operation of the sensor node, the sampling rate of ADCs and DACs must obey the Nyquist theorem. To preserve the original continuous

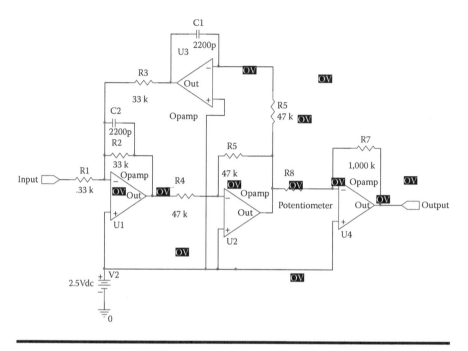

Figure 11.7 Receiver amplifier.

waveform, the sampling rate must be greater than twice the highest frequency component in the waveform. In this project, the highest potential frequency used is 25 kHz. This means that a sampling rate greater than 50 kHz would be necessary to correctly capture the frequency of the waveform without aliasing. The DSP selected for this project has onboard ADC with sampling rates up to 100 kHz.

11.2.1.6 Signal Level Detection

We also need circuitry to detect the voltage level of the incoming signal. This circuit is made up of three elements: a diode rectifier, an RC (Resister-Capacitor Circuit) low-pass filter, and a TTL (Transistor–Transistor Logic) AND gate (see Figure 11.8).

The circuit operates by first rectifying the signal. The output of this operation is essentially the absolute value of the input wave (see Figure 11.9).

After being rectified, the signal is low-pass filtered. A low-pass filter acts as an averaging filter once it reaches steady state, and at this point the signal is essentially a DC voltage level. The filtered signal is used as the input to a TTL AND gate, with the other gate input tied to V_{DD}. When the signal reaches sufficient magnitude to pass the logic "1" threshold of the AND gate, the output of the AND gate will become a logic "1," and notify the DSP that a potential signal arrived and the demodulation process should begin.

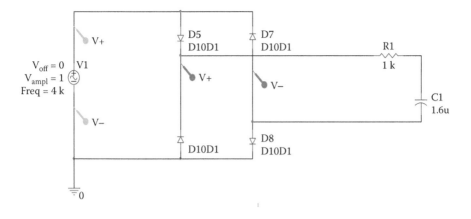

Figure 11.8 Threshold detection circuit.

11.2.1.7 Digital Signal Processor

To execute acoustic communications protocols with neighboring underwater nodes, a microcontroller is used to control all sensors. It is a fixed-point DSP with 256 kB of Direct Memory Access (DMA)-capable memory. A fixed-point processor was chosen because a floating-point processor is much more expensive, and requires substantially more power to operate. The fixed-point processor is suited towards our low-cost/low-power goal.

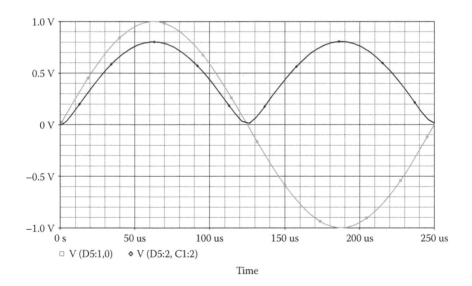

Figure 11.9 Rectification stage of threshold detector.

The microcontroller also has a substantial amount of RAM (100 kbytes), which is necessary due to the large amount of memory consumed by sampling and signal processing. The microcontroller's DMA system is also in use, which allows the sampled symbols to be placed into memory directly, instead of using valuable CPU cycles to move the samples from the ADC to memory. It allows a symbol to be demodulated concurrently with sampling, so that no samples are lost during the long demodulation calculation.

The clock frequency has been set to 14.74 MHz. At this speed, the microcontroller can fully demodulate a symbol before the next symbol is fully sampled; therefore, no samples are lost. However, for future expansion, the clock speed can be increased to as much as 80 MHz.

The microcontroller is packaged with an interface board, which contains peripherals that are directly accessible to the processor. These include a variety of digital I/O ports, a digital potentiometer-based DAC, and a 10-bit/12-bit ADC. Having these peripherals integrated directly with the CPU, our system saves substantial development time and cost.

11.2.1.8 Noise Reduction Using Adaptive Filter

The sensed pH/temperature values could have much noise from the circuit itself and the underwater environment. The above DSP has a built-in adaptive filter to reduce the data noise. The system configuration of the adaptive filter is shown in Figure 11.10. In the figure, $S(k)$ is the acoustic signal; $n(k)$ is the noise signal; $n'(k)$ is the reference signal that is being used to reduce the noise; $Sys(k)$ is the primary input to the filter that is the corrupt signal; $H(z)$ is the transfer function of the adaptive filter; $Y(k)$ is the output of the adaptive filter; and $c(k)$ is the output signal (it is the difference of the corrupt signal and the output of the adaptive filter). The reference input $n'(k)$ is correlated with the noise and is uncorrelated with the original signal. It can be proved that $n(k)$ and $n'(k)$ are correlated but $n'(k)$ and $s(k)$ are uncorrelated (see the derivation given in [11]).

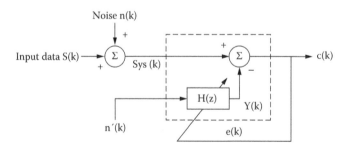

Figure 11.10 Adaptive filter configuration.

The Least-Mean-Square (LMS) filtering algorithm is used to reduce the noise since it is the simplest adaptive algorithm [12]. This algorithm uses a special estimate of the gradient that is valid for the adaptive linear combiner. It does not require off-line gradient estimations or repetitions of data. If the adaptive system is an adaptive linear combiner, and if the input vector and the desired response are available at each iteration of the algorithm, the LMS algorithm is generally the best choice for many different applications of adaptive signal processing.

11.2.1.9 Underwater Sensor Board

In summary, the underwater node we designed includes the following components: (1) microcontroller with intelligent data processing and wireless communication capabilities; (2) analog sensors to collect underwater parameters such as pH values; (3) acoustic transmitter and receiver; (4) batteries, and others. Figure 11.11 shows our fabricated underwater board (it has DSP and acoustic communication modules; analog sensors are not shown).

11.2.2 Software Design

Through modular software components the node can be easily reconfigured for different future tasks (such as ad hoc routing protocols). The software of the system is divided into two sets of components: those associated with transmission and those associated with reception. Figure 11.12 shows our software structure in the receiver side. It includes analog sensor data filtering and acoustic demodulation. Figure 11.13 is the transmitter's software structure that has modulation and CRC (Cyclic Redundancy Check) error control.

Figure 11.11 Fabricated underwater node with microcontroller and acoustic transceiver.

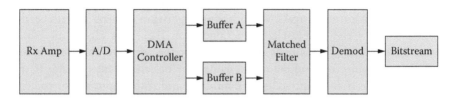

Figure 11.12   Receiver side software block diagram.

11.2.2.1 Formation of Data Packets

Data is then taken from the sensor ADCs and written directly into the data portion of the packet space. With the current sensor configuration, this stores a 16-bit integer. The fields contained in this 16-bit packet header are shown in Table 11.1.

The preamble is used by the receiving node to validate that the bit stream it has detected is indeed a packet originating from this system. The preamble is set to $0 \times A5(165)$.

The destination has two options: either the *surface node* or the *underwater node*. However, the 8-bit field allows up to 255 additional nodes to be added. The *surface node* is the node that is connected to the data center with software GUI (Graphical User Interface), and the *underwater nodes* send their data in a hop-to-hop way until finally reaching the *surface node* (see Figure 11.1).

The sequence number is increased as each packet is sent, and in the case of sending data that spread across multiple packets, the sequence numbers would be used to reassemble the data in the correct order.

The command bits are used to tell the destination node how to interpret the data contained in the packet. Up to 16 commands are supported. For instance, two frequently used commands are "sending a data packet" or "acknowledging a data packet."

Finally, the CRC field contains information to verify that the data has arrived uncorrupted. The CRC of the data is calculated on the sending side and inserted into the packet. When the packet is received, the receiver recalculates the CRC of the data, and compares it to the CRC contained in the header. If they match, it means that the data are not corrupted.

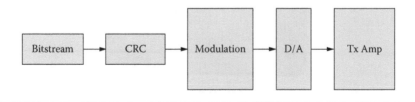

Figure 11.13   Transmitter side software block diagram.

Table 11.1   Packet Header Fields

8 bits	4 bits	4 bits	4 bits	4 bits
Preamble	Destination	Sequence number	Command	CRC

11.2.2.2 Modulation

The modulation scheme chosen for this project was FSK modulation. FSK works by changing the frequency of the carrier depending upon which binary symbol is represented. If a 2-level FSK is used, there are only two carriers. In this scenario each carrier represents only a single bit. A more complex system might use several different carriers, to represent multiple bits at a single time. For instance, four carriers could represent two bits, where the presence of any given carrier would represent a specific combination of two bits ("00," "01," "10," "11").

In our current design, the two carriers were located at 2.5 kHz (representing a "0" bit) and 3.5 kHz (representing a "1" bit). Carriers in FSK must be spaced far enough apart so that the receiver can correctly discern between any two carriers. In this system the two carriers were placed 1 kHz apart.

Modulation is designed to be able to generate the carriers at runtime within the system. Each of the carriers is generated using the following formula:

$$c = \left\lfloor \frac{MaxVal}{2} \times \sin(2\pi f t) \right\rfloor + \frac{MaxVal}{2}, \qquad (11.1)$$

where t is a vector of times defined from 0 to the end of the period of the wave, $T = 1/f$. The resolution of this vector is defined by the sampling frequency of DAC. *MaxVal* defines the maximum amplitude for the waveform. The microprocessor's DAC has a resolution of 8 bits, so its maximum value is 255. For this system the D/A sampling frequency has been set to 8 kHz. The sampling frequency was selected because the maximum frequency to transmit is 3.5 kHz. Based on Nyquist's sampling theorem, 8 kHz was the closest sampling frequency that could be obtained on the DSP board.

The modulation is accomplished by repeating the appropriate carrier wave stored in memory multiple times to achieve the desired bit rate in the system. The maximum bit rate achieved was 16 bits/second. The system originally needed to maintain 8 bits/second to keep up with the receiver's DMA. However, additional gains were necessary to ensure no loss of time samples.

Additionally, it was assumed that the receiver would be able to detect signals in the frequencies of interest, and immediately start sampling. However, the sampling system takes time to start up. To help the receiver align each symbol to be demodulated, a chirp wave is transmitted prior to the packet's preamble.

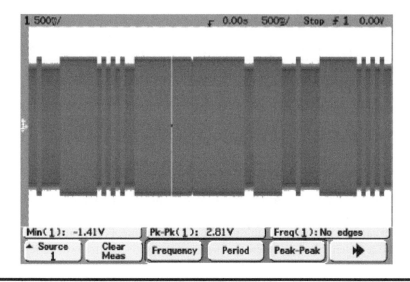

Figure 11.14 Modulation of a typical packet.

An oscilloscope capture of transmitted symbols from the modulation circuit is shown in Figure 11.14.

11.2.2.3 Packet Detection

To begin a reception, the receiving node must first detect that there is a signal present in the water. The master node runs a level detector function whenever it is idle. The level detector function takes each buffer captured by the A/D and computes the frequency spectrum. The two frequency bins for each carrier frequency used (2.5 kHz, 3.5 kHz) are compared to a threshold value. If the threshold is exceeded, then there is sufficient signal in the water to start the demodulation process.

11.2.2.4 Symbol Alignment

Once the digital signal processor receives a signal from the level detector circuit, the demodulation process begins. This starts by initiating a DMA routine capturing samples from the A/D to internal buffers, without the need for direct interaction of the digital signal processor. The DMA allows the A/D to take samples and place them in buffers without taking clock cycles away from the DSP. This allows the DSP to accomplish other computationally expensive tasks. The DSP copies the samples after a block is completed by the sampling routine. To ensure that the DSP does not miss any samples, the DMA runs two buffers in ping-pong mode. In this mode one buffer is filled with samples, and then while that buffer is being internally copied, it fills a second buffer. The DSP allocates a total of 1024 bytes for the total

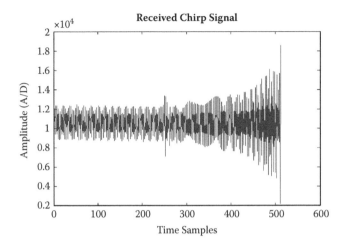

Figure 11.15   Received chirp signals.

DMA memory space. This space was simply split in half, and each buffer takes up 512 samples. Each carrier wave is also constructed of 512 samples, to match this buffer length. The A/D has a resolution of 12 bits, and is also clocked at 8 kHz.

However, the DMA takes time to initiate, and since the level-detector does not take a deterministic amount of time to detect the presence of a signal, the exact starting location of a symbol must be computed within the buffer. A chirp wave is used to help exactly compute where in any given buffer the symbol begins (an example of a delayed chirp occurring in a buffer is shown in Figure 11.15). The chirp wave is half the duration of a symbol and sweeps linearly from 2.5 kHz to 3.5 kHz. To ensure that the level-detector will activate prior to the arrival of the chirp, it is preceded by a 2.5 kHz carrier wave of the same duration.

The starting location of a chirp within the first buffer is computed using a matched filter. The reason that a chirp must be transmitted and single carrier cannot be used is because a carrier has virtually no bandwidth, and matched filters produce timely resolution (i.e., a precise location indicating the start of the chirp wave) only when given sufficient frequency bandwidth. The index into the buffer where the chirp wave begins is computed using a matched filter as follows:

$$t = \text{Max}\left(abs(IFFT(x_{\text{ref}})^* \times FFT(x)))\right), \tag{11.2}$$

where x_{ref} is the reference signal, i.e., the expected signal, in this case the chirp signal, and x is the actual received signal. The output of the matched filter function is shown in Figure 11.16. As shown, the function has a maximal output where the reference signal begins within the given buffer of samples. The delay incurred in computing

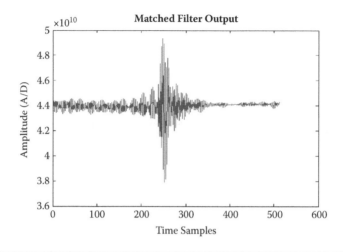

Figure 11.16   Matched filter output.

this index, however, needs to be computed only once per transmission, and introduces more robust receiving capabilities into the system for further expandability.

11.2.2.5 Demodulation

The symbol realignment index is used to determine the offset in each buffer where a symbol representing a specific bit will occur. It is necessary because a given symbol could begin two-thirds of the way through a buffer. If it is sent to the demodulator without correct alignment, the demodulator will not be able to correctly determine which signal is present in the buffer because only one third of the symbol appears in the buffer. The symbol of interest will then span two separate buffers.

Once the alignment of a symbol within the two buffers has been determined, the frequency domain representation of the symbol is computed using a 512-point FFT. Since *phase* information is not important for FSK modulation, only the *magnitude* of the FFT is needed to determine the appropriate bit. With the spectrum of the symbol computed, the two pertinent frequency bins are observed (those corresponding to each possible carrier signal). The frequency bin with highest amplitude determines which bit the symbol represents.

It is critical that the demodulation process must complete in less time than the transmission of a single symbol. The DMA sampling process is ongoing during demodulation, so the demodulation of a single bit must complete within one symbol period. Otherwise the system would not be able to run at real-time, and quickly run out of memory (and thus not be able to store buffered samples). The period of each symbol is T = 64 ms (512 × 125 uS). The demodulation routine is approximately 33 ms at 14.74 MHz clock speed, and is subsequently sufficient to run and process an entire buffer before another buffer is returned by the DMA.

11.3 System Testing

The surface node (see Figure 11.1) is connected to a PC via an RS-232 connection at 2400 baud. The surface node has no sensors of its own in this case. The underwater nodes have the pH and temperature sensors. The underwater nodes are to poll their sensors for information every 30 seconds. Once the sensor data are retrieved a packet is constructed and the CRC of the packet is computed.

After a packet is sent, it will be detected by the surface node, and then gets demodulated. The surface node will also check the CRC to make sure the packet arrived free of corruption. Next, the command field is checked to see how to interpret the accompanying data. The sensed data are to be tabulated, and to be extracted and sent to the GUI via the serial port.

The surface node must then send an acknowledgment (ACK) back to the underwater nodes so it knows the packet was received successfully and will not have to be resent. If the CRC shows that the packet is corrupted, or the packet does not arrive at all, the receiver will do nothing. Eventually the sender will time out while waiting for the acknowledgment, and send the packet again. The sequence of error recovery is shown in Figure 11.17. This is known as a stop-and-wait Automatic Repeat reQuest (ARQ) mechanism.

A lab test was conducted using these procedures. The test setup is shown in Figure 11.18. The system achieved its theoretical bit-rate of 15.625 bits/sec. The maximum number of retransmissions observed was 10. An average bit error rate (BER) of 0.091 was observed. The CRC did not avoid all errors since it is not the

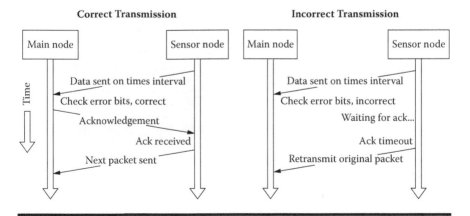

Figure 11.17   ARQ interactions.

Figure 11.18   Lab test setup.

strongest error-checking scheme in wireless networks. Some packets were deemed error-free when there was at least one bit error contained within the packet. These can be seen in Figure 11.19 where spikes are present on the GUI graph of pH temperature results. These points are obvious outliers (the pH greatly exceeds 14, and it must be a result of bit errors).

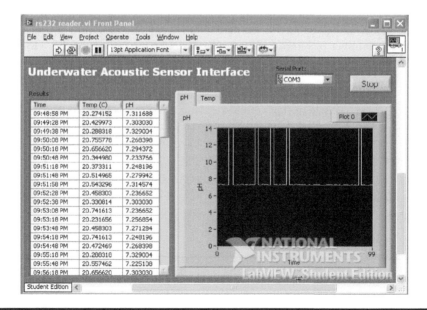

Figure 11.19   Underwater sensor data collection GUI.

11.4 Underwater Communications Security

As part of the national information infrastructure, underwater sensor networks could also become the potential network attacking objects because of their high importance to national security, public health, and ecology analysis. Through attacking underwater measuring services, adversaries will be able to cause detrimental effects on water quality monitoring. For instance, the water contaminant source cannot be accurately located; packets will be lost if the sleep-wakeup schedules among neighboring nodes are messed up (this can further trigger many unnecessary packet retransmissions if MAC layer acknowledgments are used); the water quality data cannot be correctly logged into the annual database due to no accurate routing information; and so on.

A major concern in underwater sensor network security schemes design is *energy efficiency*. Most of a sensor's battery is consumed in acoustic communications instead of in local signal processing or sensing [1]. Therefore, the security schemes should not use too many message exchanges between underwater nodes. Moreover, the security schemes should be of low complexity. Therefore, Symmetric-crypto could be a better choice than traditional Asymmetric-crypto based on public/private keys having high computational overhead.

Security in each individual hop is the prerequisite of the multi-hop sensor security. As the starting point of our security research, we implemented a low-energy, low-overhead security scheme for one-hop wireless communications [8,10]. As shown in Figure 11.20, the security software is built in both the underwater node and the interface board that serves as the transition gateway between underwater sensor networks and the Internet.

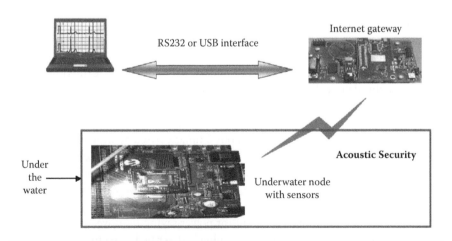

Figure 11.20 One-hop security scenario.

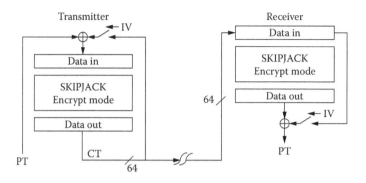

Figure 11.21 SkipJack crypto (PT, plaintext; CT, cipher text; IV, initialization vector; CBC, cipher-block chaining, which works with 64 bits).

Our underwater security mechanism uses the following two security primitives (see Figure 11.21):

1. *Initialization vectors (IVs)*: One implication of semantic security is that encrypting the same plaintext two times should give two different ciphertexts. The main purpose of IVs is to add variation to the encryption process when there is little variation in the set of messages.

2. *Block cipher choice*: Triple-DES [13] is too slow for software implementation in embedded sensors. We found RC5 [13] and SkipJack [14] to be most appropriate for embedded microcontrollers. Although RC5 is slightly faster, it is patented. Also, for good performance, RC5 requires the key schedule to be pre-computed, which uses 104 extra bytes of RAM per key. Because of these drawbacks, we selected SkipJack.

Simulation Test: It is somewhat difficult to directly measure the energy consumption of security mechanisms from a single underwater node. We have thus resorted to an accurate simulator called Power Tossim [15] where hardware peripherals (such as the acoustic transceiver, EEPROM, LEDs, and so forth) are instrumented to obtain a trace of each device's activity during the simulation run. Through the obtained real-time traces of the current drawn in our SkipJack-based Symmetric-crypto and RSA-based Symmetric-crypto [13], we have computed the energy consumption of major components (such as CPU idle, CPU active, acoustic communications, etc.) in underwater nodes (see Table 11.2). From Table 11.2, we can see that for the two most important components, i.e., CPU active and acoustic transmission, our proposed security scheme shows significant power-saving improvements over RSA (Rivest, Shamir, and Adleman, who first publicly described it) security scheme (the energy efficiency is improved by 90% and 150%, respectively).

Table 11.2 Security Energy Consumption Comparisons

Energy Consumption (in milli-joules, mJ)	SkipJack	RSA
CPU active	55	110
Acoustic communications	2198	4121
Memory access	36	81
Total	2289	4312

Lab Test: We have also tested the lifetime of the underwater nodes under the above two ciphers (i.e., SkipJack and RSA). We set up the sensing duty cycle as 70% (i.e., for 70% of the time the sensors are in active sensing status and go to sleeping status in remaining time). It is found that the same batteries can work for around 5 days under the RSA case, while they can last for over 2 weeks for SkipJack scheme. This indicates the energy efficiency of the SkipJack case. If we use duty cycle 2%, both of them can last for a long time (<1 month), and thus it is difficult to see the final lifetime. In practical sensor networks, 2% duty cycle is a reasonable setup.

11.5 Conclusions and Future Works

This research has made cost-effective underwater sensor nodes with multi-hop routing and security schemes. The hardware cost for a node is less than US$100. It is battery-driven and has reasonable lifetime (it lasts for a few months under 2% duty cycle).

This project has great potential for future expansion. Listed below are some potential areas of future work to expand this system.

The system could easily be expanded to use multiple carriers and transmit two or more bits simultaneously. Additionally, multiple-FSK (MFSK) that uses multiple simultaneous carrier waveforms could increase the communication throughput. MFSK is popular and has been used in several previous underwater systems [16,17]. Additionally, a PSK modulation scheme could be added to, or replace, the current system. The system is phase-coherent due to the matched filtering process that takes place at the beginning of the transmission. The WHOI [8] modem utilizes PSK for its high-rate data communications. PSK modulations could potentially be significantly less prone to interference, and also lower the overall system BER.

Long delay in the acoustic underwater channel creates a long contention window for ISI. The current system does not have an equalization circuit to mitigate the effects of the long channel delay. A future enhancement could greatly reduce the BER by equalizing the channel.

The amplifiers used in the system were constructed of a series of active filters. More sophisticated amplifiers could be constructed to significantly increase the range of this system. The active filters are constrained by the voltage rails provided to them, confined by the batteries used to drive the system. Use of more sophisticated amplifier design techniques as well as output/input impedance matching would help the system to reach further links, and increase the energy efficiency of the system.

Acknowledgments

The authors acknowledge the wireless communications group in Lockheed Martin Inc. (the branch in Rochester, NY) for their support of our underwater sensor network research. We also thank U.S. National Science Foundation (NSF) (Award No. 0511098) for supporting the development of the Pervasive Computing Laboratory at Rochester Institute of Technology. We also thank Cisco Inc. University Research Program (URP) for supporting the development of wireless security projects.

References

1. Deven Makhija, P. Kumaraswamy, Roy. Rajarshi "Challenges and Design of Mac Protocol for Underwater Acoustic Sensor Networks," *Fourth International Symposium on Modeling and Optimization in Mobile Ad Hoc and Wireless Networks*, April 2006.
2. John Heidemann, Wei Ye, Jack Wills, Affan Syed, Yuan Li. "Research Challenges and Applications for Underwater Sensor Networking," *IEEE Wireless Communications and Networking Conference*, April 2006.
3. Wei Ye, John Heidemann, and Deborah Estrin. "An Energy-Efficient MAC protocol for Wireless Sensor Networks." In *Proceedings of the IEEE Infocom*, pp. 1567–1576. New York, NY, USA, USC/Information Sciences Institute, IEEE. June, 2002.
4. LinkQuest. Available at: http://www.link-quest.com
5. E. Sozer, M. Stojanovic, J. Proakis. "Underwater Acoustic Networks," *IEEE Oceanic Engineering*, 2000, Vol. 25, 1.
6. DSPComm. Available at: www.dspcomm.com
7. X. Yang, K. Ong, W. Dreschel, K. Zeng, C. Mungle, C. Grimes. "Design of a Wireless Sensor Network for Long-Term, In-Situ Monitoring of an Aqueous Environment," *Sensors*, 2002, 2, 455–472.
8. Lee Freitag, M. Grund, S. Singh, J. Partan, P. Koski, K. Ball. "The WHOI Micro-Modem: An Acoustic Communications and Navigation System for Multiple Platforms," *IEEE OCEANS*, 2005, Vol. 2, pages 1086–1092.
9. R. Jurdak, C. Lopes, P. Baldi. Software Acoustic Modems for Short Range Mote-Based Underwater Sensor Networks. Available at: http://www.ics.uci.edu/~rjurdak/Oceans06.pdf
10. International Transducer Corporation. Available at: www.itc-transducer.com
11. B. Widrow, S. D. Stearns. 1985. *Adaptive Signal Processing*. New York: Prentice Hall, Inc.
12. S. D. Stearns, R. A. David. 1988. *Signal Processing Algorithms*. New York: Prentice Hall, Inc.

13. Bruce Schneier. 1996. *Applied Cryptography*, Second Edition. New York: John Wiley & Sons.
14. SkipJack specifications. Available at: http://jya.com/skip jack-spec.htm
15. V. Shnayder, et al. "Simulating the Power Consumption of Large-Scale Sensor Network Applications." In *Proc. of SenSys*, Nov. 2004.
16. K. F. Scussel, J. A. Rice, S. Merriam, "A New MFSK Acoustic Modem for Operation in Adverse Underwater Channels." In *Proceedings of MTS/IEEE Oceans 97*, Vol. 2, (Halifax, N.S., Canada), pp. 247–253, 1997.
17. M. D. Green, J. A. Rice. "Channel-Tolerant FH-MFSK Acoustic Signaling for Undersea Communications and Networks," *IEEE Journal of Oceanic Engineering*, Vol. 25, No. 1, January 2000.

Chapter 12

Channel Modeling for Underwater Acoustic Sensor Networks

Peter King, Ramachandran Vekatesan, and Cheng Li

Contents

12.1 Introduction

Reliable and efficient communication protocols form the backbone of successful underwater acoustic sensor networks. The inherent difficulty in ocean deployment, as compared to many terrestrial applications, means that development of these protocols must rely on the use of numerical analysis and simulation. For this type of development to be of most use and relevance, the underlying physical communication model that is used to represent the real-world channel must capture those aspects of the channel that are of most importance; there must exist a high level of fidelity between responses to the simulated model and that of the real-world model.

The ocean is a highly dynamic and complex environment in which to communicate. Characteristics that may represent one region of the ocean may not adequately represent another. It is this diversity that motivates our described methodology of generating a specific model for a specific region, through the use of as much real-world descriptive data as can be made available. This methodology will outline the steps in taking available data and generating a numerical representation of that environment, which can then be used for various applications. The steps will be further illustrated through the use of a case study, described in Section 12.4.

12.2 Background

12.2.1 BELLHOP

The foundations of this model are the predicted acoustic paths found using the BELLHOP beam tracing program.[1] Predicted paths are considered for a particular environment given a description of its physical properties. These properties include a profile of the speed of sound over various depths, the composition and bathymetry of the bottom, and the nature of the sound propagation above the surface. As the properties of a particular environment change, so too will the paths over which sound will travel. As properties vary from location to location, the paths will vary as well. It is this dependence on the variation of physical characteristics that motivates the use of a model that derives its foundation through the consideration of each location's unique physical characteristics.

The presented methodology utilizes the BELLHOP program as it is implemented in the Acoustic Toolbox by Michael Porter. This methodology uses a specific set of capabilities provided by BELLHOP and is not a thorough treatment on the usage of the program. Only those options that are used in this methodology are considered. For a full treatment of BELLHOP, please refer to References 1 and 2.

12.2.2 Related Work

The work by M. Stojanovic[3] offers a great description of the determination and calculation of channel capacity in a theoretical underwater environment. Her work provides much of the theory and calculations needed to establish the method for determining capacity, as well as establishing a description of bandwidth. However, the use of real ocean data for the establishment of a multipath transfer function is not considered in her work, which could be significantly different from an ideal environment. Consideration of a single-path case limits its use for practical applications.

Work by Hayward and Yang[4] also considered real data along with the BELLHOP program, and follows a similar approach as[3] to determine channel capacity. In contrast to our study, Hayward and Yang provide results for communication links in the upper half of the water column, whereas our study focuses on the near bottom case and provides multiple cases for node location. As well, our work extends the modeling to allow for integration into a simulation model for network testing.

An NS-2 model implementation by Harris and Zorzi[5] provides a great contribution to the field of network simulation. Their work provides NS-2 with a fundamental model of the underwater acoustic physical channel. Using that model, path loss and delay characteristics are generated from the well-known assumptions about acoustic propagation in the underwater channel.[6,7] Although this model provides an excellent starting point and is a great improvement over existing attempts to model the underwater channel as a special case of terrestrial radio channels, fading and multipath arrival effects are not considered.[8]

Work by Xie et al. describes a similar endeavor to incorporate a more realistic acoustic model into a simulation environment.[8] They incorporated the Monterey-Miami Parabolic Equation into the OPNET simulation tool. Unlike our proposed work, data from the analytic model is subjected to a regression prior to OPNET integration; therefore, some features of the path loss characteristics will be lost. As well, in[8] only the aggregate path loss is considered; the time-varying collection of multipath arrivals is not considered. Our work proposes to use a look-up table structure to maintain more features of the path loss and relies on the summation of arrivals over time.

12.3 Methodology

The methodology follows a number of steps from collecting real-world data to a final mathematical representation of the channel. The steps begin with the generation the environmental description file (ENV), which is required for operation of the BELLHOP program, and continue with the analysis and post-processing of the BELLHOP output, as well as the generation of the noise characteristics. This methodology has been implemented for our previous works[9,10] and is illustrated in Figure 12.1.

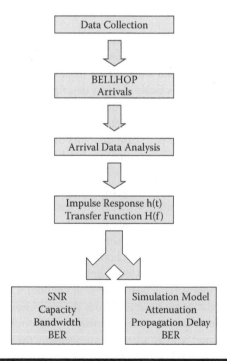

Figure 12.1 Flow diagram on steps involved in model creation.

12.3.1 Environmental File Generation

The BELLHOP program requires an input file known as an ENV. This file is the main description of the environment of interest and its fields serve as the template for providing a description of the particular physical characteristics for a given location. The ENV file is broken into sections that each requires some knowledge about the location that is to be modeled. Generation and processing of theses ENV files represents a majority of the effort that goes into the model generation. The following sections describe each step in creating the ENV files and, ultimately, in describing the environment.

12.3.1.1 Frequency Definition

The initial step in the model generation is to define the frequency at which the path calculations should be performed. Each instance of the BELLHOP program's execution allows for the definition of a single frequency; thus, to perform a sweep over a range of frequencies, a separate input file must be created for each frequency step. The range and resolution will be dependent on your particular application. The selected frequency will mostly affect the transmission loss, as absorption effects in the ocean are highly frequency dependent.[6]

12.3.1.2 SSP Description

The definition of the Sound Speed Profile (SSP) of an area of interest is perhaps the most important step in the model definition. The SSP defines the speed of sound at each depth, and allows the BELLHOP program to compute the paths that sound waves will travel through the water. Due to the nature of acoustic wave propagation, acoustic rays will not travel straight lines, but will curve toward areas of lower speed. This refraction is described by Snell's law[7] as:

$$\frac{\cos\theta}{c} = constant, \tag{12.1}$$

where θ is the angle of incidence and c is the speed of sound. The result of this is that sound rays will bend toward areas of lowest speed.

Generation of the SSP is greatly benefited through the use of actual data from the specific location of interest. This data will generally come in the form of conductivity-temperature-depth (CTD) measurements, which are often easily accessible from databases, such as those kept by the Bedford Institute of Oceanography (BIO), or the National Oceanic and Atmospheric Administration (NOAA). CTD data provides the salinity and temperature values for a series of depths, which are used to calculate the speed of sound at that particular depth using the following relationship:

$$c = 1449.2 + 4.6T - 0.055T^2 + 0.00029T^2$$
$$+ (1.34 - 0.010T)(s - 35) + 0.06z, \tag{12.2}$$

where c is the speed of sound in *m/s*, T is the temperature in Celsius, s is salinity in practical salinity units (PSU), and z is the depth in meters.[6,7]

As the temperature and salinity values vary throughout the day, month, and year, the SSP can be generated with average values for c, or multiple SSPs may be generated for use in a time- or date-specific model. Another option is to calculate mean speed values and standard deviation over a period of time. This is useful to generate a model that models the mean case and to generate models throughout the distribution of the SSP to gauge the effects of a varying SSP and to see its effect on the resultant communication channel.

12.3.1.3 Bottom Bathymetry and Description

A description of the sea floor is required to determine how the acoustic rays will interact with the bottom. A description of the bottom consists of the composition description, in terms of how sounds propagate through the bottom medium, the surface roughness, and the bathymetry for the region of interest.

Options for composition include describing the bottom half-space as a vacuum in which sound does not propagate, as a rigid surface that perfectly reflects acoustic waves, and more complicated descriptions such as an acousto-elastic half-space, in which the density and p-wave and s-wave speed and attenuation are given, or a direct description of the reflection coefficients for various angles.

The roughness is given as the root-mean-squared (RMS) value in meters of the inter-facial features of the sea bottom. This data will determine any scattering effects of the sea bottom. A separate bathymetry file (BTY) is used to describe the overall contour of the bottom and consists of a set of ranges from the transmitter and the subsequent depths. BELLHOP will fit the included bathymetry data using linear sections.

Much like the SSP, the BELLHOP program will only accept a single BTY file for each ENV file. Therefore, to fully describe an area, multiple BTY files are used, each representing a different direction. The number of directions is at the discretion of the designer, and should be chosen so as to adequately describe the region.

12.3.1.4 Source and Receiver Depths and Ranges

The BELLHOP program computes the paths between a given set of source and receiver locations. These locations are given as the depths of each source and the range and depth for each receiver. The selection of depths and ranges is directly dependent on the target application and should adequately cover the possible locations that will be occupied in the final network simulation. Since BELLHOP will compute paths for each source and receiver pair, there is a direct correlation between the number of pairs and the execution time of the program.

The values are given as a list of source depths followed by a list of receiver depths. A list of ranges at which the receivers will occupy is also given.

12.3.1.5 Run Type

The BELLHOP program has many options for the type of computations it can perform, and the output it will provide. There are three main types of output files provided by BELLHOP: Ray, Amplitude-Delay, and Acoustic Field.

Ray files plot the path of acoustic rays that propagate from each source. Options exist for describing either all rays that leave the source, option "R," or only those rays that intersect at the described receivers, option "E." Rays that intersect the receiver are known as eigenrays.

Amplitude-Delay output files list the characteristics of each eigenray in terms of the received signal strength, the time-of-arrival, phase shift, and number of reflections for both surface and bottom. Options for this type of file include binary output, "a," or ASCII output, "A."

The Acoustic Field files describe the ocean section as an acoustic field. Data of this type is not concerned with paths of propagation, but the relative signal strength, or transmission loss, at all locations within the bounding area. The method of calculation is defined by three options: Coherent option "C," Incoherent option "I," and Semi-coherent option "S."

12.3.1.6 Beam Width

To limit the number of paths that are computed, a range of transmission angles is provided to describe the width of the available beam angles. This will include the minimum and maximum angles that will be traced, where negative angles are toward the surface from horizontal and positive angles are toward the bottom. As well the number of beams that will fill this range is given. The BELLHOP program allows value of 0 for number of beams to indicate that it should automatically choose the number of beams.

12.3.1.7 Bounding Box and Step

The physical bounds of the range and depth are provided to BELLHOP to limit the extent that it traces the acoustic paths. This is the maximum limit of range and depth that will be considered in calculation. It is imperative to ensure that the bounding box is sufficiently large to encapsulate all ray paths and does not limit the calculation. The bounding box should extend slightly past the bottom depth and the maximal source range. The step size is also given to indicate the resolution at which to trace the acoustic paths. The step may be set to 0, indicating that BELLHOP should automatically determine the step size.

12.3.1.8 ENV File Template

Below is a template for the complete ENV file; values requiring user customization are indicated as "?."

'????????'	! TITLE
?.?	! FREQ (Hz)
1	! NMEDIA
'S?NT'	! Top Boundary Condition
0 0.0 ????.?	! DEPTH of bottom (m)
?.? ????.??/	! Depth SpeedOfSound
... ! this line is repeated	
?.? ????.??/	! for each pair in SSP

'?' ?.? !Bottom Option Roughness	
?	! # of source depths
????.? ..... ????.? /	! list of source depths (m)
?	! # of receiver depths
????.? ..... ????.? /	! list of receiver depths (m)
?	! # of receiver ranges
????? ????? /	! list of receiver ranges (km)
'A'	! Run-type
??	! # of beam angles
?? ?? /	! list of beam angles
??.? ??.? ??.?	! step size (m) max. depth (m) max. range (km)

12.3.2 BELLHOP Processing

Due to the nature of the BELLHOP program, computing arrivals for each SSP, bathymetry, and frequency variation cannot be accomplished by executing a single ENV file. Depending on the scope of the network being modeled, there may be hundreds or thousands of ENV files that need to be analyzed.

A script is used to handle processing of the ENV file variations. This may include pre-generation of all ENV variations and then iterative execution of the BELLHOP program. As well, certain parameters of the ENV file may be declared dynamically prior to BELLHOP iteration.

12.3.3 Arrival File Analysis

For each ENV file processed by the BELLHOP program, a resulting arrival file (ARR) is generated. This file contains information about each path that connects a transmitter position to a receiver position. A single ARR file will contain the arrivals for each pair of declared source and receiver positions.

Arrival data is given in the following format:

1.3705288E–03 90.00000 0.6263431 –8.976132 –9.047516 0 1

where the fields are

Amplitude—relative received signal strength after propagation and absorption effects have been accounted for

Phase—received signal phase in degrees

Delay—propagation time in seconds

Transmit angle—angle from transmitter beam path followed

Receive angle—angle to receiver beam path followed

Surface reflections—total number of times path has reflected off the surface

Bottom reflections—total number of times path has reflected off the seafloor

12.3.3.1 Channel Response

The ARR file data is treated as the impulse response of the communication channel between the transmitter and receiver. With this in mind, it is possible to derive the channel's frequency response.

For each pairing of transmitter and receiver and each bathymetry case, a transfer function is calculated. This transfer function is calculated through summation of each arrival. That is,

$$S_a = \Gamma e^{j\phi} \tag{12.3}$$

where S_a is a particular arrival, Γ is the amplitude, and ϕ is the phase in degrees. Γ and ϕ are taken directly from the ARR file.

The transfer function is then calculated through summation of the arrivals at each particular frequency, shown as

$$H(f) = \sum_{\alpha=0}^{A-1} S_\alpha e^{-j2\pi f r_a}, \tag{12.4}$$

where A is the total number of arrivals, H is the frequency-dependent transfer function, and T is the particular path arrival time.

12.3.4 Noise Analysis

Modeling the acoustic channel noise involves the summation of noise from various sources. Thermal effects, shipping activity, wind, and turbulence effects all contribute noise within the spectrum of acoustic communications and thus are all considered. Depending on the particular scenario and situation, noise data may be determined in one of two ways: through direct measurement of current or historic noise power spectral density, or through the use of established equations and assumptions.

12.4 Case Study

To best illustrate the steps and process of generating a region-specific model, a step-by-step case study is provided. This study is based on data collected off the east coast of Newfoundland at a location of 46 29.5N 48 29.4W. Data is taken from conductivity-temperature-depth (CTD) measurements made available by Oceans Ltd., a St. John's-based oceanographic company, and from bathymetry data taken from the Bedford Institute of Oceanography's (BIO) database.

As per Section 12.3, the case study will outline the process of creating of the environmental description file (ENV) through to the generation of the channel response function. A description of installation and configuration of the BELLHOP program is not given; the reader is encouraged to refer to the relevant resources for a treatment on platform specific installation. For reference, the case study was performed using the Acoustic Toolbox implementation of BELLHOP, as installed under Ubuntu Linux; as well, much of the post-BELLHOP processing and visualizations were performed using MATLAB®.

12.4.1 Environmental File Generation

12.4.1.1 Frequency Definition

For our case study, the frequencies we will process are in the range from 0 Hz to 60 kHz, at steps of 250 Hz. The range is based on our experience with the usable frequency range for underwater communication and the step size is selected to balance the generation of smooth curves and to allow for manageable computational duration. This selection will require that we generate 240 ENV files, one for each frequency.

12.4.1.2 Sound Speed Profile

At our location, CTD data is available for an artificially constructed 12-month period, based on data collected over a 10-year period. The data is centered about four depths, sampled at 20-minute intervals. The case study will consider one month, August, and find the average Sound Speed Profile (SSP) for that particular month. This method can be repeated for each month to derive a month-specific model that can be used throughout the year.

Data is provided as a text file with fields of day, month, year, hour, minute, salinity in practical-salinity-units (PSU), temperature in Celsius, and depth in meters. A MATLAB® script is used to filter through the file and select the temperature and salinity values for a particular month. Figure 12.2 shows the temperature and salinity values for August.

Using Equation (12.2) the speed of sound values are calculated and a shape-preserving fitting algorithm is used to generate the SSP. Figure 12.3 shows the calculated speed values and the resultant SSP.

For entry into the ENV file, a set of depth (m) and speed (m/s) pairs are extracted from the SSP. For this case study the resultant entries are given in Table 12.1.

12.4.1.3 Bottom Description and Bathymetry

In this study, no specific data is used for the description of the bottom composition. Based on discussion with those familiar with the area, the bottom is treated as a rigid surface with an inter-facial roughness of 0.2 m. Often times, not all required data will be available; in these cases a best effort assumption is necessary.

The bathymetry is taken from the National Oceanic and Atmospheric Administration's (NOAA) National Geophysical Data Center (NGDC). The data is in the form of position and depth pairs centered about 46.5N, 48.5W at 1-minute intervals. Figure 12.4 is a three-dimensional rendering of the bathymetry data for the area.

Since the BELLHOP program can only operate in two dimensions, we must create multiple ENV files to process the bathymetry in all directions from the center. For the case study, eight slices of the bathymetry were used representing directions of E, SE, S, SW, W, NW, N, NE from the center. The resultant bathymetry (BTY) file entries are given in Table 12.2.

12.4.1.4 Source and Receiver Depths and Ranges

The case study is based on the development of a sea floor sensor network. This application determines the positions we are interested in analyzing in our model. For this study, the node transducers will be positioned at a depth of 5 m above the sea floor; thus, the transmitter will be at a depth of 90 m, since the area's center depth is 95 m. The depths of the receiving nodes will vary depending on the bathymetry, which will give a range of depths from 87 m to 92 m. The nominal horizontal distances will be 1 km and 2 km, so to allow for variance, ranges of 0.9 km, 1 km, 1.1 km, 1.9 km, 2 km, and 2.1 km are used.

Entry into the ENV file is as follows:

```
1 ! # of source depths
90.0 / ! source depth (m)
```

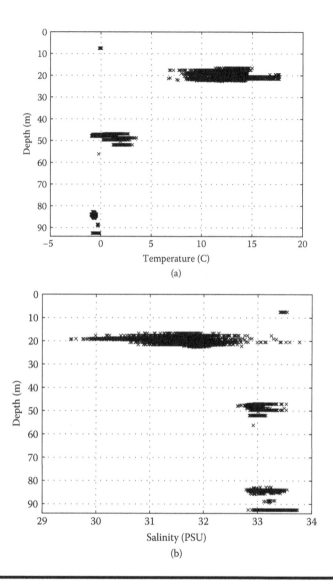

Figure 12.2 Raw data for a 1-month period. (a) Temperature. (b) Salinity.

6 ! # of receiver depths
87 88 89 90 91 92/ ! receiver depths (m)
6 ! # of receiver ranges
0.9 1.0 1.1 1.9 2.0 2.1 / ! receiver ranges (km)

It is important to ensure that the selected source and receiver positions allow proper coverage for your particular application.

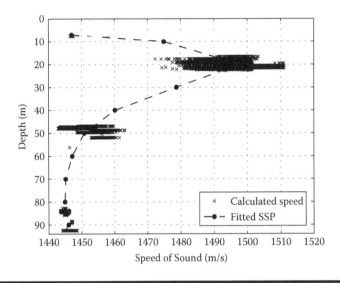

Figure 12.3   Calculated speed values (m/s) and resultant Sound Speed Profile.

Table 12.1   Sound Speed Profile

Depth (m)	Speed (m/s)
7	1447.0
10	1474.9
20	1497.7
30	1478.7
40	1460.1
50	1450.8
60	1447.1
70	1445.1
80	1444.9
90	1446.0
95	1443.8

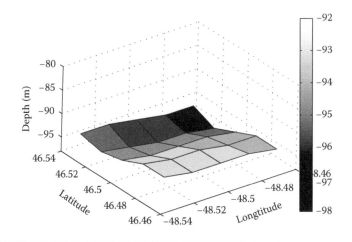

Figure 12.4 **Bathymetry around the case-study region, centered at 46.5N, 48.5W.**

Table 12.2 **Bathymetry Values**

E		SE		S	
Range (km)	Depth (m)	Range (km)	Depth (m)	Range (km)	Depth (m)
0	95.0	0	95.0	0	95.0
1.276	95.0	2.25	94.0	1.853	93.0
2.551	96.0	4.5	94.0	3.706	93.0
SW		W		NW	
Range (km)	Depth (m)	Range (km)	Depth (m)	Range (km)	Depth (m)
0	95.0	0	95.0	0	95.0
2.25	93.0	1.276	95.0	2.25	96.0
4.5	92.0	2.551	94.0	4.5	94.0
N		NE			
Range (km)	Depth (m)	Range (km)	Depth (m)		
0	95.0	0	95.0		
1.853	97.0	2.25	97.0		
3.706	95.0	4.5	96.0		

12.4.1.5 Run Type

For the development of our model, we will use the Amplitude-Delay file format, given as option "A." This will provide a set of arrival paths for each combination of source and receiver locations. The determined amplitude, delay, and phase shift for each path will be used in the creation of the channel response.

12.4.1.6 Beam Width

For this case study, a beam width of 13° is chosen from −10° to +3°, with the number of beams set at 0. Setting the number of angles to 0 informs BELLHOP to select a value automatically. The beam angles were chosen based on initial tests that indicated that no arrivals were generated outside of this range. In the BELLHOP program negative angles are toward the surface while positive angles are toward the bottom.

```
0 ! # of beams
−10.0 3.0 / ! minAngle maxAngle (degrees)
```

12.4.1.7 Bounding Box and Step

The case study has node depths that range from 92 m to 97 m. To ensure full coverage the vertical boundary is given as 100 m. The horizontal boundary is set at 2.2 km, as the longest distance required is 2.1 km. It is suggested that the boundaries slightly extend past the largest values used to ensure proper calculation. The step size is selected as 0 to allow BELLHOP to chose an appropriate value.

```
0 100 2.2 ! step (m), depth (m), range (km)
```

12.4.1.8 Completed ENV File

Below is a completed ENV file for a 1 kHz run.

```
'UWASN'! TITLE
1000.0 ! FREQ (Hz)
1 ! NMEDIA
'CVFT'! SSPOPT
0 0 95 ! DEPTH of bottom (m)
7 1447/
10 1474.9/
20 1497.7/
30 1478.7/
```

```
40 1460.1/
50 1450.8/
60 1447.1/
70 1445.1/
80 1444.9/
90 1446.0/
95 1443.8/
'R*' 0.02
1 ! NSD
90.0 / ! SD(1:NSD) (m)
6 ! NRD
87 88 89 90 91 92/ ! RD(1:NRD) (m)
6 ! NR
0.9 1.0 1.1 1.9 2.0 2.1 / ! R(1:NR) (km)
'A' ! 'R/C/I/S'
0 ! NBeams
−10.0 3.0 / ! ALPHA1,2 (degrees)
0 100 2.2 ! STEP (m), ZBOX (m), RBOX (km)
```

12.4.2 BELLHOP Processing

At this point in the process there will exist a set of ENV files, each representing a description of the environment through a specific SSP, frequency, or BTY file. The number of ENV files will depend on your application, and the amount of variability required.

The case study example includes frequency variations from 0 to 60 kHz, in increments of 250 Hz, for a total of 240 frequencies. As well, there are eight bathymetry cases from the center of the region to each of the directions selected for analysis. We are using a single SSP to represent the average profile for the month of August; therefore, we require eight sets of 240 ENV files representing the frequency spread for each direction, for a total of 1920 ENV files.

To provide a level of automation in the process, some simple scripts are used to perform the iteration of BELLHOP on each of the individual ENV cases. For each bathymetry case, a simple directory structure is used to organize the files and allow the processing to take place. An "INPUT" directory holds a collection of ENV files, each representing a single frequency. A simple naming convention of "f.env" is used, such that "1000.env" would represent the ENV file for the 1000-Hz case.

To take advantage of our computer's dual-core processor we utilize two temporary directories, "TMP1" and "TMP2," which allow two instances of BELLHOP to operate concurrently. An "OUTPUT" directory is used to store the resultant arrival data files (ARR). At the root of the bathymetry director are the relevant BTY file and the script used to perform the iterative processing. Table 12.3 illustrates the directory structure.

Table 12.3 Directory Structure

TopDir/			
	BtyN/		
		N.bty	
		MultiRun.sh	
		INPUT/	
			0.env
			1.env
			…
			60000.env
		TMP1/	
			Run.sh
		TMP2/	
			Run.sh
		OUTPUT/	

Processing is handled by a shell script, named here as *MultiRun*. The script processes each ENV file by sequentially copying each ENV file into one of the two TMP directories. It also copies over the BTY file, renaming it to match the ENV file; this is required as BELLHOP uses the ENV filename to find the BTY file. The *MuliRun* script will then call the *Run* script, which initiates BELLHOP for the particular ENV filename. Upon BELLHOP's completion, the resultant ARR files are copied into the OUTPUT directory, and the process continues for the next ENV files.

Below is a listing of the *MultiRun* script. As stated, it is designed to allow two instances of BELLHOP to run concurrently; for single or quadcore processors it can be modified.

```
COUNTER=0
let MAX=60000
while [ $COUNTER -lt $MAX ];
do
#launch on CPU-1
cp ./INPUT/$COUNTER.env ./TMP1/
cp ./Enviro.bty ./TMP1/
cd TMP1
```

```
mv Enviro.bty $COUNTER.bty
./RunBellhop.sh $COUNTER &
B1=$!
cd ..
#increment by frequency step
let COUNTER2=COUNTER+250
#launch on CPU-2
cp ./INPUT/$COUNTER2.env ./TMP2/
cp ./Enviro.bty ./TMP2/
cd TMP2
mv Enviro.bty $COUNTER2.bty
./RunBellhop.sh $COUNTER2 &
B2=$!
cd ..
#wait for both to complete
wait $B1
wait $B2
#fetch output files
mv ./TMP1/$COUNTER.arr ./OUTPUT/
rm ./TMP1/$COUNTER.*
mv ./TMP2/$COUNTER2.arr ./OUTPUT/
rm ./TMP2/$COUNTER2.*
let COUNTER=COUNTER2+250
done
```

For our setup, files were processed two at a time on a dual-core Xeon T5400 3.33 GHz computer with 2 GB of RAM. Execution of the files ranged from 0.128 s for the lowest frequency to 186 s for the highest frequency, for a total execution time of approximately 25 hours. Depending on the number of transmitting and receiver nodes, as well as values for step and beams, this value can vary greatly. It should be noted that this step is a one-time-only pre-processing step and will not need to be repeated for future use of the resultant model.

12.4.3 Arrival Analysis and Channel Response

Once collected, the collection of ARR files each represents the impulse responses for a particular frequency, bathymety, and SSP. With this data the channel's frequency-dependent response can be determined as described in Section 12.3.3.1.

If we consider a single bathymetry direction and a single pairing of source and receiver, we determine the channel response through summation of each arrival over each frequency. Seen in pseudocode this is shown as

```
for f = 0:250
H[f] = 0.0
```

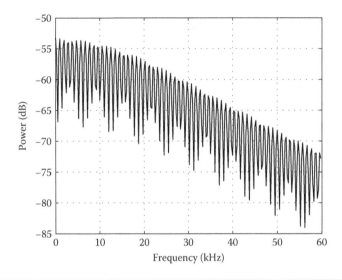

Figure 12.5 Transfer function H over frequency for single bathymetry and transmitter-receiver pair.

```
for a = 1:NumArrivals
H[f] = H[f] + A[a] * exp(-i*2*pi*Freq[f]*delay[a])
end
end
```

where *f* is the index through each of the frequencies, *H*[] is the resultant channel response, a is the index through the arrival, bounded by *NumArrivals*, *A*[] is the amplitude of the particular arrival, *freq*[] is the frequency value at the index *f*, and *delay*[] is the time delay of the particular arrival, relative to the first arrival.

Figure 12.5 shows the transfer function over frequency for a single bathymetry and transmitter-receiver pair.

The set of frequency responses for the channel represent the model of the channel. Section 12.5 will describe applications that use this information, along with the noise analysis described in the next section.

12.4.4 Noise Analysis

For the case study, no direct measured noise data was available so well-known noise equations are employed that use numeric descriptions of wind speed and a shipping constant to determine the overall frequency-dependent noise factor, as given in[3] as

$$N(f) = N_t(f) + N_w(f) + N_s(f) + N_{th}(f) \qquad (12.5)$$

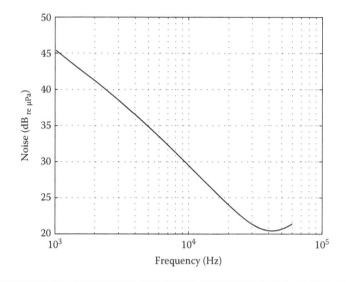

Figure 12.6   Noise power spectral density.

where $N(f)$ is the total frequency-dependent noise power spectral density (psd), $N_t(f)$ represents noise due to turbulence, $N_w(f)$ is wind noise based on the wind speed, $N_s(f)$ is noise from shipping activity and is calculated from an activity factor ranging from 0 to 1, and $N_{th}(f)$ is the noise from thermal effects.

The noise psd for the frequency range of 0 to 60 kHz is shown in Figure 12.6. Further explanation of the noise calculations can be found in[3].

12.5 Application

12.5.1 *Numerical Channel Model*

Obtaining a set of channel responses and noise characteristics provides the basis for other numerical considerations of a communication channel. These include the bandwidth and the theoretical channel capacity. As well, the channel response provides the means for performing signal analysis on coding and modulation schemes that may be considered for underwater communication.

12.5.1.1 *Bandwidth*

Channel bandwidth is an important description of how a communication channel will perform. The bandwidth is centered on a center frequency, at which the relationship between the arrival strength and the noise is at a maximum. From this maximal point, the bandwidth can be extended to include frequencies at which the signal-to-noise ratio is within some set threshold, such as 3 dB. We calculate the bandwidth through the relationship $H(f)/N(f)$. We then calculate the bandwidth for each transmitter-receiver pair, over each of the defined bathymetries.

The calculation of bandwidth is given in the following pseudo-code:

```
Fc = 0;
for f = 0 to Fmax
S(f) = H(f) / N(f);
if S(f) > S(Fc)
Fc = f;
end
...
```

where Fc is the center frequency, $S(f)$ is the signal-to-noise ratio at frequency f, and $H(f)$ and $N(f)$ are the transfer function and noise, respectively, at f.

To calculate the bandwidth, we first define the threshold. For example a 3 dB bandwidth would include those signal-to-noise values that are greater than half of the power of the center frequency. The following pseudo-code is used to compute the channel bandwidth:

```
i=0;
d=2;
for f = 0 to Fmax
S(f) = H(f) / N(f);
if S(f) > S(Fc) / d
BW[i] = f;
i++;
end
bandwidth = BW[i-1] – BW[0];
...
```

where BW becomes an array of frequencies, *Fmax* is the maximum frequency considered, d is the factor used to define the boundary for inclusion in the *bandwidth*, and bandwidth is the resultant bandwidth. To extend this beyond the 3 dB case, other values of d would be used.

From the case study we examine one of the H/N relationships in Figure 12.7, where the bandwidth is found to be 40 kHz.

12.5.1.2 Channel Capacity

Channel capacity is calculated to determine the maximum theoretical bit rate that can be expected for a particular channel given the source power, channel transfer

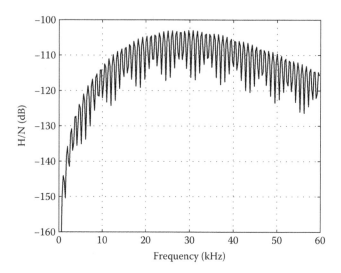

Figure 12.7 Frequency response of channel.

function, and noise power spectral density. This maximal limit is calculated by way of the equation

$$C = \sum_{i} \Delta f \log_2 \left[1 + \frac{X(f)H(f)}{N(f)} \right] \qquad (12.6)$$

where $X(f)$ is the transmitted signal power spectral density (psd), $H(f)$ is the channel transfer function, and $N(f)$ is the noise psd. Since the transfer functions are calculated for discrete frequency steps, the capacity contribution for each frequency step, Δf, is summed.

For the case study the transmitter psd, $X(f)$, is given in Figure 12.8. Data on the psd of a transducer can be found either from the manufacturer or through in-house testing.

12.5.1.3 Received Signal Analysis

Given the channel transfer function it is possible to determine the output signal for any given input signal. This is useful for analyzing various coding and modulation schemes that may be used in a particular underwater acoustic channel, as well as performing simulation on equalization techniques. Application of this would simply involve numerically performing the convolution of the described input signal with the determined transfer functions.

Figure 12.9 shows a standard MATLAB output for a random integer signal, QAM modulated, passed through an AWGN channel. Figure 12.10 is the same

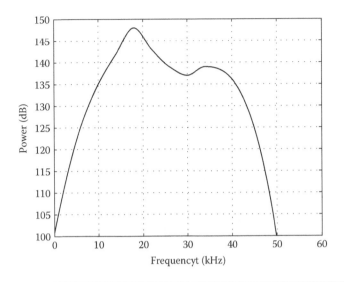

Figure 12.8   Power spectral density of transmitting transducer.

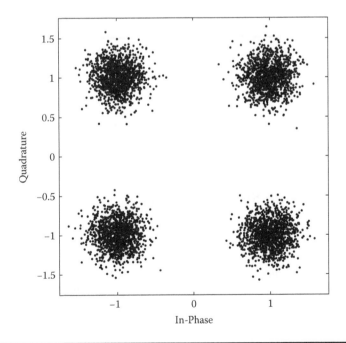

Figure 12.9   Random bit QAM signal through AWGN channel.

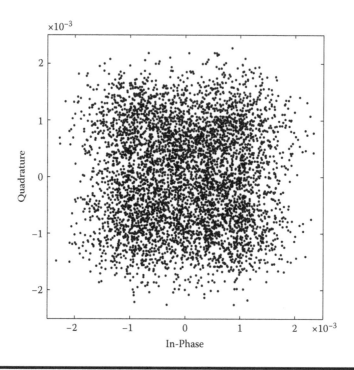

Figure 12.10   Random bit QAM signal through AWGN channel, convolved with channel response from case study.

signal after convolution with a derived channel response from the case study. The increased disorder in Figure 12.10 is due to the inclusion of the multipath effects from our derived channel model.

12.5.2 Comment: Simulation Model

Due to the inherent difficulty in deployment, there is a strong reliance on simulation for any communication system that is bound for the ocean environment. Testing the performance and robustness of protocols involves implementation in some form of simulation environment. Often this will take the form of a software-based discreet time simulation package, such as NS-2 or OPNET.

Software simulation can never fully capture all aspects of the real environment, but efforts can be made to provide fidelity to an acceptable degree. The inclusion of a physical layer model that is based on real-world measurements would certainly be a step in the right direction. With our determined transfer functions and noise profiles, it is possible to implement a set of functions to be included in the simulator that would satisfy our need for an accurate representation, and also the need for computational efficiency required by a simulator. Simulations rely on the ability to perform many repetitions and iterations to expose trends in the performance; a model that requires too much time too compute could inhibit the ability to gain enough run time to expose such trends.

For implementation into a simulator, a physical model must provide certain functionality. These functions include the propagation delay between nodes, the received signal strength from a given transmitter, the effective range of transmission, the channel noise, and the available bandwidth. In terms of the case study, focus has been on the NS-2 simulation package.

12.5.2.1 Propagation Delay

For each communication event between nodes, the simulator will request the propagation delay from the physical model to determine when a transmission should be scheduled for reception by the receiving node. In fulfilling this request the physical model will have knowledge of the positions of the nodes and can retrieve the arrival time from a simple look-up structure. Since our previous processing has included iterative runs for each node position and bathymetry, it is a matter of searching a table of arrival delays for the appropriate value. Such a structure is easily implementable in software and does not incur a large performance hit.

In the event that inadequate values exist for range and position indices, it is possible to either return the delay for the closest index, or to perform simple regressions to estimate the delay.

From our case-study data the earliest arrival times can be stored for each of the bathymetry-location-depth sets. This data can be indexed based on all parameters, or averaged and indexed solely by range, or range-depth. Below is a table comprising range indexed propagation delays.

```
Range Delay
0.9 tau
1.0 tau
1.1 tau
1.9 tau
2.0 tau
2.1 tau
delay = GetPDelay(TxPos, RxPos) {
dX = RxPos.X - TxPos.X;
dY = RxPos.Y - TxPos.Y;
dZ = RxPos.Z - TxPos.Z;
dist = sqrt(X*X + Y*Y + Z*Z);
delay = PDelayTable[dist];
return delay;
}
```

12.5.2.2 Received Signal Strength

Of most importance to a packet simulator is the knowledge of the received signal strength for each transmission. This value is used to determine if a packet is received or not, and if so, what probability of errors are associated with it. Again a look-up table can be used to index pre-calculated values based on the positions of the transmitting and receiving nodes. The values inserted in the table are calculated from the transfer function derived from the BELLHOP model. The received signal power becomes the product of the transmitted power multiplied by the complex transfer function value at the center frequency, since that will represent the strongest received strength.

```
s = GetRxStrength(TxPos, RxPos, TxPow) {
H = ChannelResp[TxPos, RxPos];
s = H * TxPow;
return s;
}
```

12.5.2.3 Effective Range

When a transmitting node sends a signal, it is important to know which nodes will be affected by its transmission, whether they are intended listeners or not. The effective range defines the range at which a transmission will be detected based on its transmission power P_t, and the sensitivity of receiving nodes P_r. In essence, this function must determine at what distance the transmission's attenuation will not exceed $P_t - P_r$. Nodes within this distance will be aware of transmissions from the particular node.

Since our case study has finite node positions, the distances that are considered are also finite.

12.6 Summary

With the highly dynamic nature of the ocean, and the vast variance between locations, it is imperative that models of the underwater acoustic channel be tailored to each individual situation. The methodology presented here provides a basis for establishing models that are built upon the characteristics that define each location.

Wherever possible, data from the environment should be used, as this will help ensure that the model is a true reflection of the region. When not available, care should be taken in the use of assumptions. When data is known to vary, this variance should also be considered in the modeling, either through variance in the parameters, or as multiple iterations of the model. The end result is a set of frequency-dependent channel responses that have application in describing the communication channel in terms of its capacity, bandwidth, and error rate prediction.

References

1. M. Porter. Ocean acoustics library. Online (December, 2007). URL: http://oalib.hlsre-search.com
2. O. Rodriguez. *General Description of the BELLHOP Ray Tracing Program (October 2007 release)*. Universidade do Algarve, 1.0 edition (2007).
3. M. Stojanovic. On the relationship between capacity and distance in an underwater acoustic communication channel, *SIGMOBILE Mob. Comput. Commun. Rev.* 11(4), 34-43, (2007). ISSN 1559-1662. doi: http://doi.acm.org/10.1145/1347364.1347373.
4. T. J. Hayward and T. Yang. Underwater acoustic communication channel capacity: A simulation study. In *Proc. of the HF Ocean Acoustics Conference*, La Jolla, California (March, 2004).
5. I. A. Harris and M. Zorzi. Modeling the underwater acoustic channel in NS2. In *Proc of ValueTools 07: The 2nd international conference on Performance evaluation methodologies and tools* (2007).
6. R. J. Urick. *Principles of Underwater Sound*. (New York: McGraw-Hill, 1983), 3rd edition.
7. Y. L. Leonid Brekhovskikh. *Fundamentals of Ocean Acoustics*. (Berlin: Springer-Verlag, 1982).
8. L. Diaz-Gonzales, G. Xie, and J. Gibson. Incorporating realistic acoustic propagation models in simulation of underwater acoustic networks: A statistical approach. In *Proc. of the Oceans' 06 MTS/IEEE, Boston* (2006).
9. P. King, R. Venkatesan, and C. Li. A study of channel capacity for a seabed underwater acoustic sensor network. In *Proc. of the Oceans' 08 MTS/IEEE*, Quebec City, PQ (September, 2008).
10. P. King, R. Venkatesan, and C. Li. An improved communications model for underwater sensor networks. In *Proc. of the IEEE GLOBECOM 2008*, New Orleans, LA (December, 2008).

Index